What is M

In What is Mind? Abhijit Naskar, bestselling author and one of the world's celebrated neuroscientists offers a fascinating account of the cellular building blocks of mind. He boldly reveals, Neuron is to Mind, what Gene is to Life. With a researcher's flair for fresh approaches to ancient questions, Naskar tackles the most controversial problem in the history of philosophy - how physical processes in the brain give rise to our lavishly colored mental lives enriched with ecstasies and agonies?

Also by Abhijit Naskar

The Art of Neuroscience in Everything

Your Own Neuron: A Tour of Your Psychic Brain

The God Parasite: Revelation of Neuroscience

The Spirituality Engine

Love Sutra: The Neuroscientific Manual of Love

Neurosutra: The Abhijit Naskar Collection

Homo: A Brief History of Consciousness

Autobiography of God: Biopsy of A Cognitive Reality

Biopsy of Religions: Neuroanalysis towards Universal Tolerance

Prescription: Treating India's Soul

NASKAR

What is Mind?

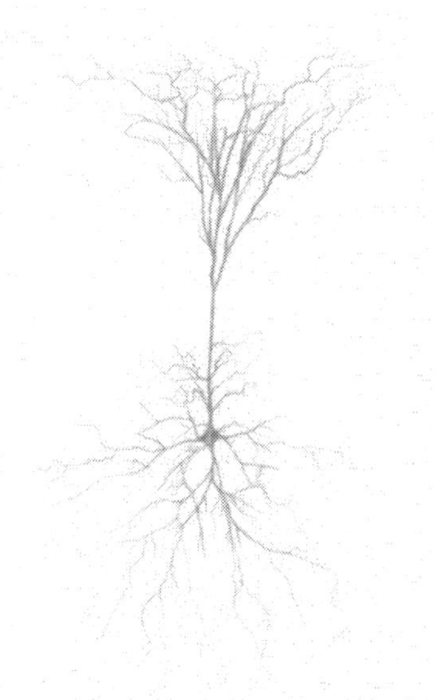

What is Mind?

This is a work of non-fiction

An Amazon Publishing Company, 1st Edition, 2016

Printed in United States of America

ISBN-13: 978-1535080552

In Memoriam

Santiago Ramón y Cajal (1852-1934)

"As long as our brain is a mystery, the universe, the reflection of the structure of the brain will also be a mystery"

Contents

Acknowledgments

A book that attempts to solve the apparently insoluble problem of the history of philosophy, can never be a creation of one scientist. Here, I may have given several fresh scientific approaches towards solving the hard problems of human mind, that have tormented the philosophers for millennia, but I – *Naskar* is the product of almost a century of rigorous neuroscientific studies and experiments.

I am simply the expression of Neuroscience itself.

And behind this scientist who does nothing but pursue complicated problems of the human mind with a naïve perspective, there are many scientific minds and one non-scientific mind at play. And, not mentioning them would only make me feel that the book is not complete in some way.

Michael A. Persinger is the first one to mention because, without him I might have always remained ignorant of the astounding Science of the Brain. Until I came across his work, I barely knew anything about the term *Neuroscience*. He made me realize that it is no other field of Science but Neuroscience that holds the key to solving the quintessential problems of life. He coaxed me into the science of the neurons and the rest is history. Without Persinger, Naskar and Neuroscience would never have been linked together.

Then there are Charles Darwin, Roger Penrose, and Erwin Schrodinger. And my dear friends Andrew Newberg, V.S. Ramachandran and Ronald Cicurel whose unique perspective of the mind have been immensely helpful.

And the one childishly non-scientific mind who molded a childish brain into an excessively curious one is Gadadhar Chatterjee. Without this man, I'd have remained a rat in the race. Without this man, Naskar would have never been truly awakened from the deep sleep of ancient ignorance. He didn't have even the basic literacy to understand the English alphabets, yet to me he's The Philosopher of All Philosophers, and the Thinker of All Thinkers. *A thousand Platos, Socrateses and Descarteses would have to merge, for one Gadadhar to be born.*

* * *

Preface

We humans cherish ourselves to be the smartest species on earth. Our vanity is in our uniqueness. Our vanity is in our unpredictability. Our vanity is in our rich, vivid and unique mental lives. Hence, nothing is more chastening to that vanity than the very realization that the richness and vividness of our mental universe, with all our personal thoughts, feelings, emotions and the very sense of our intimate selves, arises exclusively from the activity of little wisps of protoplasm in the brain.

Every time, we Neuroscientists utter *"we are but a bunch of neurons"*, the very vanity of the human species somehow gets aggravated. We take immense pride in our uniquely advanced human consciousness. Hence, the very thought that, the Advanced and Unique Human Consciousness is after all the product of various complex physiological processes, is frightening to many, especially the philosophers.

For a long time, the philosophers have been keeping the domain of understanding the human consciousness a glaring mystery due to their own lack of comprehension into the matter. For their personal inability to understand consciousness, they have been referring to it as the hard problem. After all, philosophy may enable you to think in a different way, but it is the method of Science, that enables us to understand the underlying mechanism of every phenomenon of this universe.

The so-called hard problem of consciousness is the question of how physical processes in the brain give rise to subjective experience. A similar problem used to haunt every biologist until about half a century ago, i.e. what is life? And the fascinating discovery of the structure of DNA by James Watson and Francis Crick in 1953 put an end to that hard problem of life.

This discovery was something so enormous, that it revolutionized biology, giving it an intellectual framework for understanding how information from the genes controls the functioning of the cell. That discovery led to a basic understanding of how genes are regulated, how they give rise to the proteins that determine the functioning of cells, and how development turns genes and proteins on and off to determine the body plan of an organism. With these extraordinary accomplishments behind it, biology assumed a central position in the constellation of sciences, one in parallel with physics and chemistry.

And the person who put the first stepping stone in the path towards solving the hard problem of consciousness, was the great Spanish anatomist Santiago Ramón y Cajal. He formulated the neuron doctrine, the basis for all modern thinking about the nervous system.

Cajal was the first one in our line of Neuroscientists. And we have come a long way since his formulation of the neuron doctrine. Rhetorically speaking, in the field of Neuroscience, we have advanced from canoes to galleys to steamships to space shuttles. And our relentless venture into the human brain has led us to the conclusion, that – **Neuron is to the Mind, what Gene is to Life.**

Neurons are the building blocks of mind. And in this book you shall witness how the physical processes of these building blocks construct every human's unique, individualistic and rich mental life.

I have written this book in the purpose of explaining the biological origin of mind for the readers from all walks of life. The science of the mind not only gives us insight into how we perceive, learn, remember, feel and behave – but also it gives us new perspective of ourselves in the context of biological evolution. It makes us appreciate every single biological element that works harmoniously with each other to construct something so fascinating as the human mind. Therefore, the sole purpose of this book is to contribute to the well-being of the society, by making the biology of mind accessible to every single person.

* * *

CHAPTER 1

Neuron

Human Mind - it all begins with the Neuron - about a hundred billion of them – sending and receiving electrochemical signals among each other, thus constructing a complicated mesh of inexplicable features. Just imagine, this very network is the birthplace of all your emotions, ambitions, conscious awareness and experiences. This network made you fall head over heels in love when you had your first crush at school. This network made you cherish the taste of human lips when you had your first kiss. This network made you feel special when you had your first computer. Every single human experience that you can think of right at this very moment, has been the product of electrochemical impulses running through the byzantine web of nerve cells or neurons inside your brain. In this chapter I present an empirically fresh approach towards understanding the Mind. Through the investigation of the activity of nerve cells throughout the entire nervous system, I propose that our perception of every single element of our mental life - or simply the Human Mind - is the collective expression of a hundred billion nerve cells working harmoniously with each other. This very expression is composed of various cognitive features, that ultimately reflect our environment through us, and in the process constructs our illusory realm of the mind. The Human Mind is merely a biological mirroring mechanism, that mirrors our external world.

Introduction

"The Human Brain" as most of us neuroscientists keep seeking opportunities to say out loud "is the most complexly organized structure in the universe". To appreciate this you just have to look at some numbers. The brain is made up of one hundred billion neurons, which form the basic structural and functional units of the nervous system. Each neuron makes something like one thousand to ten thousand connections with other neurons and these connecting points are known as synapses. It is here in the synapses, that exchange of information occurs, that ultimately gives rise to your colorful mental life.

Based on this information, it has been calculated that the number of possible permutations and combinations of brain activity, or in other words, the number of brain states, exceeds the number of elementary particles in the known universe. Even though it is common knowledge in our field of Neuroscience, it never ceases to amaze me that everything we experience is simply the activity of these little specks of jelly in our heads. Our entire mental life is born from relentless physical processes of these neurons. Neurons are the building blocks of our mind. There is nothing else.

Neuron – The Building Block of Mind

Given this staggering complexity, one must start with the anatomy of the very biological unit of the mind - the Neuron.

The biology of neurons is founded upon three principles:

1. The Neuron Doctrine
2. The Ionic Hypothesis
3. The Chemical Theory of Synaptic Transmission

All these three principles emerged for the most part during the first half of the twentieth century. And till this day they construct the very core of our understanding of the brain's functional organization.

The Neuron Doctrine, which many of us Neuroscientists like to call *The Cell Theory of The Brain,* states that the nerve cell or neuron is the fundamental building block and elementary signaling unit of the brain.

The Ionic Hypothesis focuses on the transmission of information within the nerve cell. It describes the mechanisms whereby individual nerve cells generate electrical signals, called action potentials, that can propagate over a considerable distance within a given nerve cell.

The Chemical Theory of Synaptic Transmission focuses on the transmission of information between nerve cells. It describes how one nerve cell communicates with another by releasing a chemical signal called a neurotransmitter. The second cell recognizes the signal and responds by means of a specific molecule in its surface membrane called a receptor.

All these three principles focus on individual neurons. And when we bring billions of neurons together they construct the nervous system. You can think of the neuron as a miniature self-contained information processor. It receives inputs, processes information, and generates outputs. The structure most associated with receiving is called the dendrite, the structure most associated with processing is called the cell body, or soma, and the structure most associated with the output is known as axon.

Mind – Nature's Own Mirror

Like any other cell, a neuron or a nerve cell has a cell body wrapped inside a cell membrane. It has a nucleus that contains the chromosomes which constitute the genetic information. The nucleus can range from 3 to 18 micrometers in diameter. The cell body also has other standard cellular components such as mitochondria, golgi bodies, nissil bodies, endoplasmic reticulum and so on. But what distinguishes a neuron from most other cells is the rich and elaborate wiring (axon and dendrite) that emerges from the cell body.

The wiring conveys the electrical message, whereas the cell body, or to be more specific the nucleus of the cell body, is where the processing of information occurs. This way, you can think of the nucleus as the brain of the neuron. In the nucleus many important cellular activities are initiated.

Based on the functional characteristics of the neurons, we can classify them into three types – sensory neuron, motor neuron and interneuron.

Neurons that directly or indirectly carry electrical signals or messages from the outside world to the spinal cord and brain are called sensory neurons.

Neurons that carry messages to the outside world in order to control the movement of muscles and activities of glands are called motor neurons.

Other than these two specialized groups of neurons, there is a large group of neurons which do not form connections with sensory receptors or muscles or glands, but just with other neurons. They are called interneurons.

Now let me elaborate a little more on the functioning of these three types of neurons.

Sensory neurons convey various sensory information such as sight, taste, smell, touch, hearing etc. from the outside world to the inside domain of human mind. These nerve cells of the nervous system are responsible for converting external stimuli from the environment into internal electrical impulses.

Motor neurons carry electrical signal from the spinal cord to the effector organs, mainly muscles and glands. Cell body of a motor neuron is located in the spinal cord and its axon that consists of efferent nerve fibers projects outside the spinal cord to directly or indirectly control the effectors.

The interneurons act as middlemen between sensory or motor neurons and the central nervous system (brain and the spinal cord).

The exclusive features of all neurons are the processing and transmission of cellular signals. Given their diversity of functions performed in different parts of the nervous system, there is, as expected, a wide variety in their shape as well as size. For instance, the cell body of a neuron can vary from 4 to 100 micrometers in diameter.

Neurons also come in various shapes. However, the typical neuron has multiple dendritic projections and one axon from the cell body. This is called a multipolar neuron.

But there are also neurons that have only one dendritic projection and these are called bipolar neurons, and some that only have one projection that includes both the dendrite and axon and these are called unipolar. To some extent the shape represents the function. For example unipolar and bipolar neurons are more typically sensory neurons, while multipolar neurons are more typically motor or interneurons.

On a closer look, you can distinguish two distinct portions in the wiring system of the structure of a typical neuron: one portion has shorter, more densely distributed wiring, known as the dendrites, the other portion, consisting typically of a single long wire, known as an axon, branches out into smaller axon terminals at the far end. Neurons use these dendrites and axons to receive and transmit signals to each other.

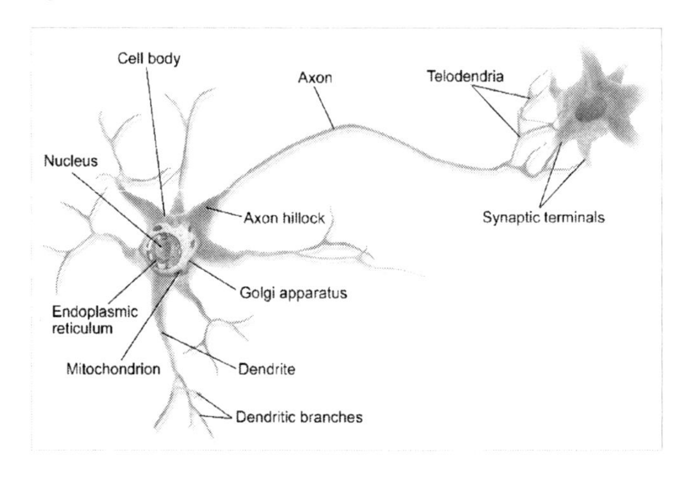

Figure 1.1 A Typical Neuron

Signals from other neurons are received by the dendrites, while signals to other neurons are transmitted by axon and its terminals. Thus in a neural wiring the dendrites act as the inputs and the axon terminals as its outputs. Signals from one neuron to another are transmitted across a small gap, between the axon terminal of one neuron and the dendrite of another neuron, known as the synapse.

A vital feature of the axon terminals is that they contain the chemical messengers, known as neurotransmitters, that are responsible for communication of neurons. And this fascinating communication among neurons gives rise to our mental universe.

Santiago Ramón y Cajal – The First Neuroscientist

Now let me tell you about the person who made it possible for us to study mental life at a cellular level. This person was, the Spanish anatomist Santiago Ramón y Cajal. He laid the foundation for the modern study of the nervous system and is arguably the most important brain scientist who ever lived. He had originally aspired to be a painter. To become familiar with the human body, he studied anatomy with his father, a surgeon, who taught him by using bones unearthed from an ancient cemetery. A fascination with these skeletal remains ultimately led Cajal from painting to anatomy, and then specifically to the anatomy of the brain. In turning to the brain, he was driven by the same curiosity that almost more than a century later drove me. Cajal wanted to understand the cellular processes of the mind.

Figure 1.2 Santiago Ramón y Cajal (1852-1934), the great Spanish neuroanatomist, formulated the neuron doctrine that is the basis for all modern thinking about the nervous system. (Courtesy of the Cajal Institute.)

He thought, the first step was to have detailed knowledge of the cellular anatomy of the brain. This is what Freud wanted to do in his early career and being unable to do so, he moved on to a more abstract field of what was then called science of the mind. Freud didn't make enough efforts to have a detailed understanding of the brain, hence he ended up cooking up the theory of psychoanalysis. Cajal on the other hand wanted to develop a true "rational psychology", which his contemporary Sigmund Freud, was unable to do.

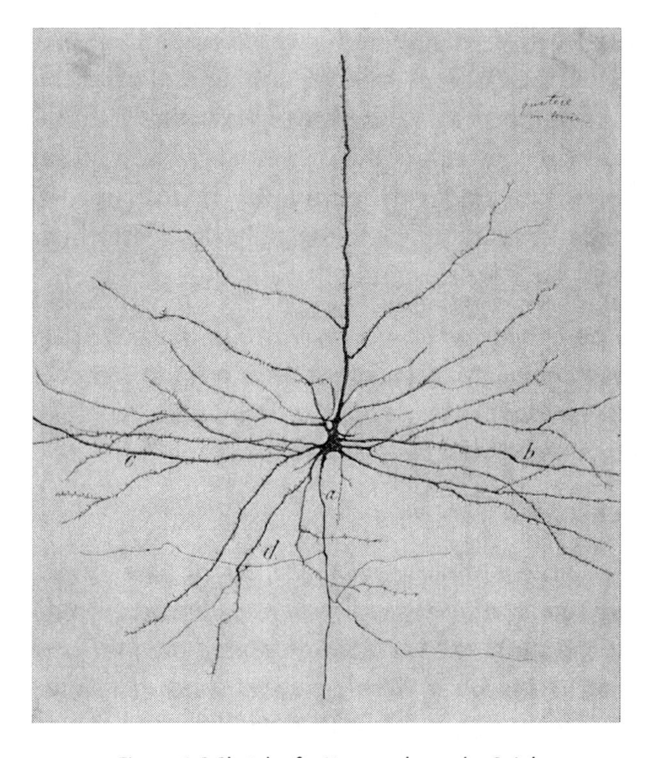

Figure 1.3 Sketch of a Neuron drawn by Cajal

Cajal was a brilliant blend of artistic and scientific genius. And this strange fusion enabled him to visualize the scenes under his microscope, as if they were vividly alive. He brought to his task of developing a "rational psychology" that very bizarre ability to infer the properties of living nerve cells from static images of dead nerve cells.

This leap of imagination, was founded upon his artistic knack. And this very imagination enabled him to capture and describe in vivid terms through beautiful drawings the essential nature of any observation he made under the microscope. The noted British physiologist Charles Sherrington would later write of him:

"in describing what the microscope showed, (Cajal) spoke habitually as though it were a living scene. This was perhaps the more striking because ... his preparations (were) all dead and fixed."

Sherrington went on to say:

"The intense anthropomorphic descriptions of what Cajal saw in stained fixed sections of the brain were at first too startling to accept. He treated the microscopic scene as though it were alive and were inhabited by beings which felt and did and hoped and tried as we do. ... A nerve cell by its emergent fiber "groped to find another"! . . . Listening to him, I asked myself how far this capacity for anthropomorphizing might not contribute to his success as an investigator. I never met anyone else in whom it was so marked."

Prior to Cajal's entry into the field, biologists were thoroughly confused by the shape of nerve cells. Unlike most other cells of the body which have a simple shape, nerve cells have highly irregular shapes and are surrounded by a multitude of exceedingly fine extensions known at that time as processes. Biologists did not know whether those processes were part of the nerve cell or not, because there was no way of tracing them back to one cell body or forward to another and thus no way of knowing where they came from or where they led. In addition, because the processes are extremely thin (about one-hundredth the thickness of a human hair), no one could see and resolve their surface membrane. This led many biologists, including the great Italian anatomist Camillo Golgi, to conclude that the processes lack a surface membrane. Moreover, because the processes surrounding one nerve cell come in close apposition to the processes surrounding other nerve cells, it appeared to Golgi that the cytoplasm inside the processes intermingles freely, creating a continuously connected nerve net much like the web of a spider, in which signals can be sent in all directions at once. Therefore, Golgi argued, the fundamental unit of the nervous system must be the freely communicating nerve net, not the single nerve cell.

In the 1890s Cajal tried to find a better way to visualize the nerve cell in its entirety. He did so by combining two research strategies. The first was to study the brain in newborn rather than adult animals. In newborns, the number of nerve cells is small, the cells are packed less densely, and the processes are shorter. This enabled Cajal to see single trees in the cellular forest of the brain. The second strategy was to use a specialized silver staining method developed by Golgi. The method is quite capricious and marks, on a fairly random basis, only an occasional neuron—less than 1 percent of the total number. But each neuron that is labeled is labeled in its entirety, permitting the viewer to see the nerve cell body and all the processes. In the newborn brain, the occasionally labeled cell stood out in the unlabeled forest like a lighted Christmas tree. Thus Cajal wrote:

"Since the full grown forest turns out to be impenetrable and indefinable, why not revert to the study of the young wood, in the nursery stage, as we might say? ... If the stage of development is well chosen . . . the nerve cells, which are still relatively small, stand out complete in each section; the terminal ramifications . . . are depicted with the utmost clearness."

These two strategies revealed that, despite their complex shape, nerve cells are single, coherent entities. The fine processes surrounding them are not independent but emanate directly from the cell body. Moreover, the entire nerve cell, including the processes, is fully enclosed by a surface membrane, consistent with the cell theory. Cajal went on to distinguish two sorts of processes, axons and dendrites. He named this three-component (cell body, axon and dendrites) view of the nerve cell the neuron. With rare exceptions, all nerve cells in the nervous system have a cell body that contains a nucleus, a single axon, and many fine dendrites.

The Big Picture

Now let's speed things up a notch and look at the big picture - the nervous system. Activity of billions of neurons throughout the entire nervous system gives rise to the functional product of the psychological system known as the human mind.

The nervous system has three main functions: sensory input, processing of data and motor output. Sensory input is when the body gathers information or data, by way of neurons, glia and synapses. The system is composed of excitable neurons. The neurons convey electrical impulses from sensory receptors to the brain and spinal cord. The data is then processed in various neural regions of the brain. After the brain has processed the information, impulses are then conveyed from the brain and spinal cord to muscles and glands, which is called motor output.

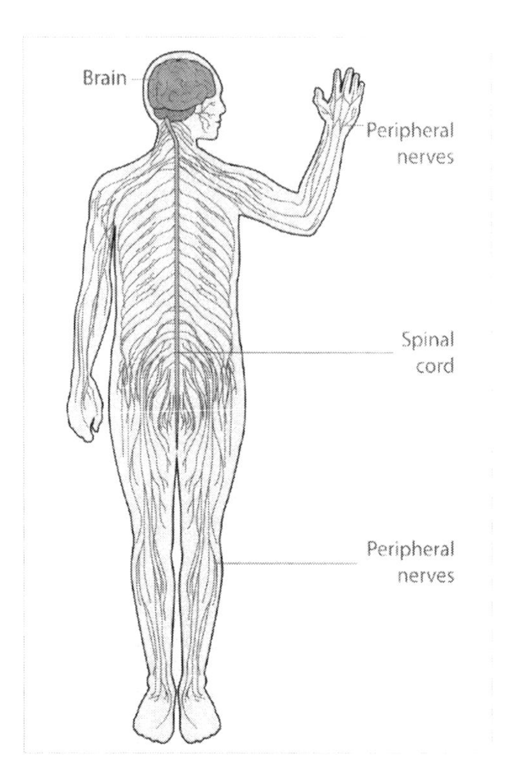

Figure 1.4 The Central and Peripheral Nervous Systems. The central nervous system, which consists of the brain and the spinal cord, is bilaterally symmetrical. The spinal cord receives sensory information from the skin through bundles of long axons that innervate the skin. These bundles are called peripheral nerves. The spinal cord also sends motor commands to muscles through the axons of the motor neurons. These sensory receptors and motor axons are part of the peripheral nervous system.

Other than the nerve cells there is another type of cells that are found in the brain and the spine.

These are called "Glia cells". The glial cells are not excitable, but they surround the neurons and provide support for and insulation between them. These cells are the most abundant cell types in the central nervous system.

The nervous system is comprised of two major parts, or subdivisions: the central nervous system (CNS) and the peripheral nervous system (PNS). The CNS includes the brain and spinal cord. The brain is the body's control center. The CNS has various centers located within it that carry out the sensory and motor processing of data.

The Peripheral Nervous System or PNS is a vast network of spinal and cranial nerves that are linked to the brain and the spinal cord. It contains sensory receptors which help in processing changes in the internal and external environment. This information is sent to the CNS via afferent nerve fibers. The PNS is subdivided into the autonomic nervous system and the somatic nervous system.

The autonomic system handles all the involuntary control of internal organs, blood vessels and cardiac muscles. These are the biological processes that go on inside your body without your voluntary involvement and conscious awareness. The somatic system handles all the voluntary control of skin, bones, joints, and skeletal muscle.

These two systems function together, by means of nerves from the PNS entering and becoming part of the CNS, and vice versa.

The Central Nervous System and the Peripheral Nervous System together have a fundamental role in everything you experience, do or feel. They construct your entire neurobiology, which is the birthplace of your very perception of being alive and being yourself. And all of this happens through the transmission of electrical signals in the tiny wonders, that we call neurons.

The Giant Input-Output System

Our entire neurobiology acts as a giant input-output system (comprised of billions of tiny input-output systems, known as the neurons), that receives information from the outside world, processes that information and makes a person react accordingly. The physical processes of the neurons do not create anything completely on their own, rather they merely react according to the outside world. This way, our entire perception of the outside world is only a reflection of the things we can physically perceive with our senses. And if we turn off all the input channels of our neurobiology there will be no way for us to respond to the outside world because we cannot sense it in the first place.

The evolutionary cause behind any living organism's ability to respond to its environment is to sense the changes in the environment and thereby adapt to it by a suitable biological response, in the ultimate pursuit of survival. The same goes for the most advanced living organism on this planet – the Homo sapiens.

This basic ability to interact with the environment is what the philosophers so mysterious venerate as Consciousness. It is a fundamental element of the human mind, and an exclusive possession of the neurons. No other species on earth may possess a beautiful and inexplicable mind like ours, but every living species possesses consciousness. Consciousness is a crucial element of sustaining life. In fact, Life and Consciousness go hand in hand. One cannot survive without the other. Without the presence of consciousness, life doesn't exist. Without consciousness we won't exist.

Without our sensory perceptions we cannot survive on our own even for a few days. One way or another our inability to interact with the environment, shall lead us to our death. I'll illustrate this scenario with a thought experiment.

The Isolation Conundrum (Thought Experiment)

Let's imagine, in a near future, our planet is hit by a medium sized asteroid. The asteroid itself is not the problem here. Rather, it possesses a strange virus, that is going to cause a mass extinction event on the planet by damaging all the sensory regions of the human brain. Soon it spreads throughout the globe in the form of an epidemic. Slowly it causes the entire humankind to lose all sensory perceptions one by one.

Humans begin to lose their senses one at a time. Each loss is preceded by heightened activity in the emotional center of the brain, the amygdala, which leads to various intense emotions right before the loss of each sensory perception.

First, people begin to suffer from an uncontrollable urge to cry, that is soon followed by the loss of their sense of smell. Then an outbreak of irrational panic and anxiety is followed by the loss of taste. Now as people have lost two of their senses their brains try to adapt to the loss. The brains attempt to fight back the loss of two senses by generating more neural connections to strengthen rest of the sensory perceptions through the process of neuroplasticity.

However, the epidemic does not stop there. Loss of hearing comes next, followed by an outbreak of extreme rage. Like what any other living organism would do, humans struggle to adjust and to go on living, with an intensified sensory perception of vision and touch. Then one day people feel an extreme sense of fear, and immediately after that the whole world goes blind.

Now the only available sensory perception of the human mind, is the sense of touch. Even when you are blind and deaf, you can still distinguish humans from the animals or any other element of the nature, by touching them. You can even tell whether it is a child or an adult. You can also tell whether it is a male or a female. So, there is still a way for you to survive by depending on your one sensory perception. If you are a young adult, you can still find another consensual young adult of the opposite sex, to mate with. You can gather food and water, by recognizing them in the nature with your sense of touch. So, with that one means to interact with your environment you still have a diminished form of consciousness. You'd struggle at first, but in time you'd get used to survive as primitive human beings only upon the reliance of one sensory perception.

Now comes the worst and final hit of the epidemic. One day, every human being on planet earth experiences a feeling of extreme euphoria. And followed by this global euphoria, humankind loses its last trace of sensory perception – the sense of touch. And with this loss of the only remaining sensory perception of the human species, you become completely isolated from the environment. Now you have literally no way of communicating with your external world. Just imagine, you don't have any means to recognize even a single element of your external world, because the physical senses that construct the mind's perception of the elements in your surroundings are not working. With the use of your motor control of the muscles, you extend your hand a little. But, because you have no sense of touch, you don't have any idea whether you are touching anything at all or not, let alone recognizing what you are touching. If not due to accidents, (like falling off a roof, falling into a body of water, etc.) you'd soon die out of hunger and thirst, because you are neurologically incapable of interacting with the environment. It means, you are also incapable of gathering food and water.

As a result, due to the lack conscious awareness of the external world, all humans on planet earth begin to die one after another. And the whole planet witnesses a mass extinction event, caused solely by the elimination of the neural ability to interact with the environment.

This little experiment makes us realize the significance of the neurobiology's input-output system. The neurobiology of our species, simply receives information from our environment, processes it, and reacts to that information.

Thus, I deduce that all our unique human achievements may seem to be the creation of the unique human mind, but they are simply the by-products of our neurobiological response to the environment. All these by-products collectively construct the human mind. And the more complex our neurology becomes over time, the more by-products shall be born out of the fascinating functioning of a hundred billion nerve cells. Thus, as the neurobiology of our species gets more and more complicated, its functional expression, which we call the mind, shall get more vivid and productive.

However, without the presence of an environmental stimulus the fantastic network of our neurobiology is completely incapable of constructing any kind of emotions, thoughts or perception, which by definition are the unique characteristics of the human mind. Thus, just by taking away the brain's ability to receive information from the external environment, we can cripple most of the human mind. However, even without the presence of sensory stimulation from the external world, the mind still possesses few crucial traces of consciousness. These traces are the thought processes and imaginations that can still go on inside the neural network by the internal stimuli of previously received environmental information. And, along with the external stimuli, by any superficial means, if we turn the internal stimuli of previously received environmental information off, then the so-called Mind would become completely non-existent.

Conclusion

It's like we act as a mirror for the environment, or to a broader aspect, the universe. The universe perceives itself through us, or to be more specific, through our neurons. Thus we can say that, each one of us is a miniature version of the infinite universe out there. We reflect every single element of our surrounding, and in the process we feel unique and special in doing so. Because we are completely unaware of the entire process. We are completely unaware that all through our lifetime, we act nothing but at a relay system for information. We convey message through space and time. And in return we receive this extraordinary sensation of being alive – an inexplicable sensation of possessing an unparalleled Human Mind.

* * *

CHAPTER II

Brain

An entire life, lavishly colored with ecstasies and agonies, takes place in a three pounds lump of jelly, which we call the Brain. It is an organ with unparalleled importance. Just imagine, everything that makes you who you are, is born from this remarkable neural cauliflower. The neural circuits of this spongy material define your uniqueness. It defines your emotions, dreams, behaviors, personality, and even your religious sentiments.

Introduction

When you are saying "I", it is actually the billions of neurons in your brain collectively expressing their functional existence. *Soul* is nothing but a bunch of neurons firing relentlessly. Everything that makes you, you, is a biologically existential expression of your entire brain. We can even say that *"you are your brain"*, and *"your brain is you."*

Cerebral Hemispheres

Now, let's start with some basic anatomy of this magnificent organic structure. In this twenty-first century most people have a rough idea of what the brain looks like. It has two mirror-image haves, called the cerebral hemispheres, and resembles a walnut sitting on top of a stalk, called brainstem. Each hemisphere is divided into four lobes: the frontal lobe, the parietal lobe, the occipital lobe, and the temporal lobe.

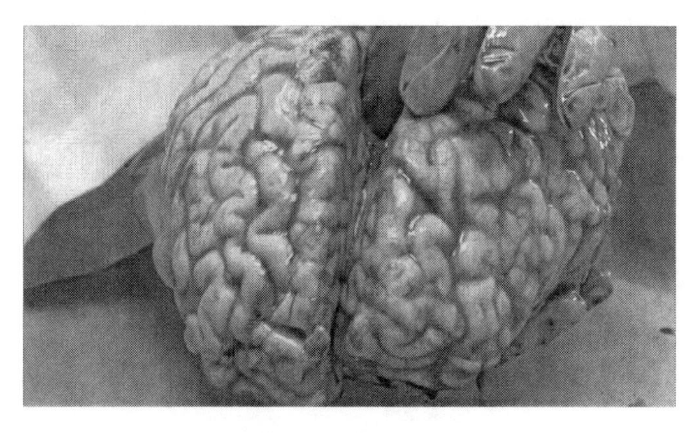

Figure 2.1 Two Hemispheres of the Brain

The occipital lobe at the back of your head is concerned with vision. Damage to this region can result in blindness. The temporal lobe is responsible for mostly hearing, and certain aspects of visual perceptions, along with the unique human feature of religious sentiments. Anomalous activity in the neural network of this region, often leads to experiences that people tend to define as religious or spiritual. The parietal lobe is concerned with constructing your perception of the external world. It simulates a three dimensional representation of the spatial layout of the environment you live in, and also of your own body within that three dimensional representation. And lastly the frontal lobes, are perhaps the most enigmatic of all the brain regions. They are concerned with some very exuberant aspects of the human mind, such as your moral sense, your wisdom, your ambition and other mysterious activities of the mind, which we have only started to understand.

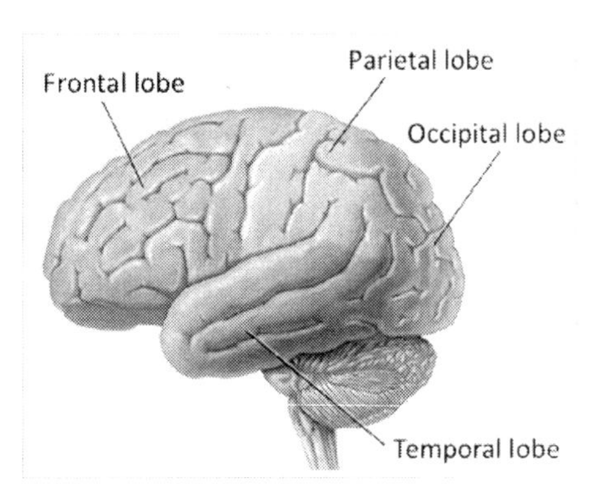

Figure 2.2 Four Lobes of the Brain

The cerebral hemispheres contain higher nerve centers responsible for various sensory and motor information. Whereas the brain stem contains neural networks that constitute lower nerve centers for the control of vital functions of the body such as breathing, blood pressure regulation, sleeping, eating, heart rate monitoring etc. To put it simply, the brain stem modulates vital processes of your body, without your conscious awareness of involvement. Because, consciousness itself is the functional expression of mostly the higher nerve centers residing in the cerebral hemispheres.

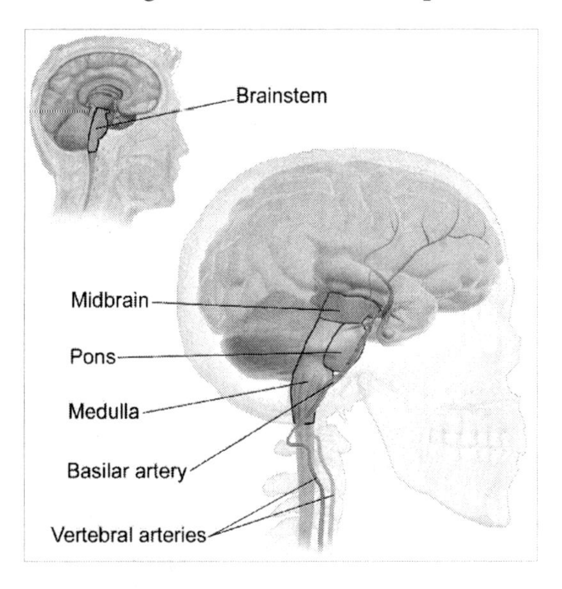

Figure 2.3 The Brain Stem consists of three parts – the midbrain, medulla, and pons

The brain stem is divided into three distinct parts: medulla oblongata, pons and midbrain.

The midbrain contains groups of neurons that project up to the cerebral hemispheres. These neurons modulate bodily functions like alertness, temperature regulation, pleasure etc.

The pons lies between the medulla oblongata and the midbrain. It contains tracts that carry signals from the cerebral hemispheres to the medulla and to the cerebellum and also tracts that carry sensory signals to the thalamus.

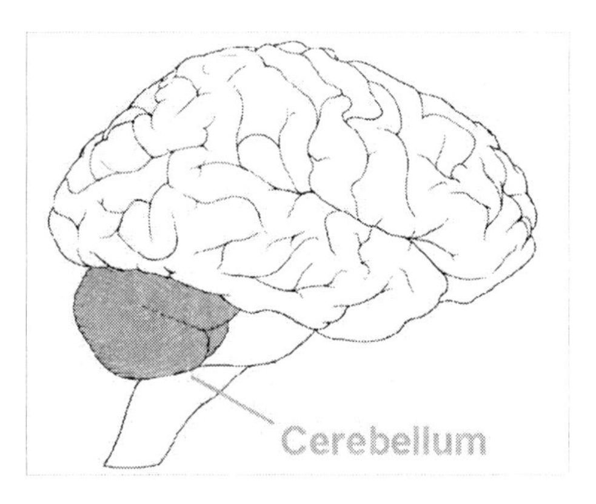

Figure 2.4 Cerebellum provides you with smooth, coordinated body movement

Cerebellum is a beautiful neuronal machine of the brain whose intricate cellular architecture plays an absolutely central role in the control and timing of movements. It receives input from sensory systems of the spinal cord and from other parts of the brain, and integrates these inputs to fine-tune motor activity or muscle movement. Damage to this little part of the brain leads to poorly coordinated movements, loss of balance, slurred speech, cognitive difficulties and also motor learning disability. The cerebellum is vital for motor learning and adaptation.

Almost all of your voluntary actions rely on fine control of motor circuits, and the cerebellum is important in their optimal adjustment, with respect to timing. It has a very regular cortical arrangement and seems to have evolved to bring together vast amounts of information from the sensory systems, the cortical motor areas, the spinal cord and the brainstem.

Imagine yourself playing catch. If you are a pro player, then it apparently seems so easy to catch the ball coming towards you. But underneath that simple action of catching the ball, lies intricate neural functioning of the cerebellum. It carries out countless electrochemical tasks inside its architecture with the help of higher nerve centers of the brain, to enable you to perceive the whole situation - the position of the ball, its speed, its distance from you, the projectile of its motion. And the end product of such fine tuning of neural functioning, is you catching the ball with a blink of an eye.

So, the next time when you play, remember to appreciate the fascinating functioning of your brain before you catch the ball. Remember to realize that almost all levels of your motor hierarchy are involved in that apparently simple task - from planning the action in relation to the moving visual target, programming the movements of your limbs, and adjusting the postural reflexes of your arm. At all stages, you would need to integrate sensory information into the stream of signals leading to your muscles.

Every time, I think about these magnificent biological processes, it leaves me in a feeling of awe. This is what we can call one of the most fascinating features of the human mind. It is plain beautiful.

Now before we move on to the brain structures containing higher nerve centers, let me give you a brief introduction to the third part of the brainstem. While talking about the midbrain and pons, we shifted our attention a little towards the brain region called cerebellum, the very neural architecture that is fine tuning your body posture right at this very moment, while you are reading this book. It is making sure that you are comfortable enough to pay attention to the pages, that are explaining to you its very functioning.

Now, the third part of the brainstem is known as medulla oblongata, or just medulla. It is the lower half of the brainstem continuous with the spinal cord. Its upper part is continuous with the pons. Right in front of the cerebellum, the medulla is cone-shaped neural mass that connects the higher levels of the brain to the spinal cord.

It is responsible for several functions of the autonomous nervous system:

1. Respiration - Medulla modulates respiration process with the use of chemoreceptors that transduce a chemical signal into an action potential. These receptors detect changes in acidity of the blood, thus if the blood is considered too acidic by the medulla, electrical signals are sent to intercostal muscles of the chest cavity, increasing their contraction rate in order to reoxygenate the blood,
2. Cardiac functioning - Heart rate monitoring.
3. Vasomotor functioning - Regulation of blood pressure with the use of baroreceptors, that sense blood pressure,
4. Modulating reflex actions of vomiting, coughing, sneezing, and swallowing. These reflexes which include the pharyngeal reflex,

the swallowing reflex (also known as the palatal reflex), and the masseter reflex, can be termed, bulbar reflexes.

In short medulla oblongata plays a crucial role in maintaining homeostasis in your body.

Now let's go deeper into the brain to discover the structures that together construct all the emotional features of our mental life.

The Limbic System – House of Emotions

Right above the brainstem, there is a system of inexplicable properties. It is the very heart of all your emotions. It is the complex neural architecture, that combines higher mental functions and emotional responses into one system. I am talking about a fascinating part of the brain, known as the Limbic System. Let me give you a little example to show the significance of this system in human life.

When you say to someone "follow your heart", it actually refers to the rhetorical representation of various emotions, that are precisely produced from neural activity of the limbic system. So, the metaphoric heart we always boast about while giving advice to our friends, is actually not anywhere near the biological organ known as heart. Rather it too, like all other elements of the human mind exists only in the brain.

This is why we often refer to the limbic system as the emotional nervous system.

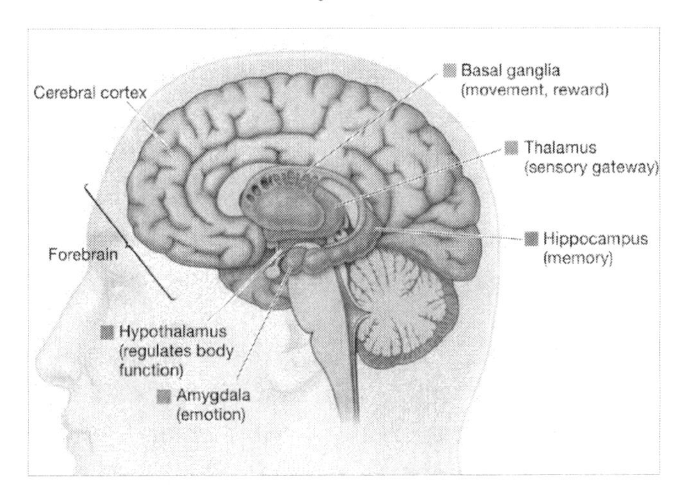

Figure 2.5 The Limbic System surrounded by the layers of Cerebral Cortex

The Limbic System is a complex set of structures found just beneath the cerebrum. It is not only responsible for our rich and colorful emotional lives, but also some of our higher mental functions, such as learning and formation of memories. The limbic system is composed of several structures, each of which plays crucial role in simulating our mental lives. They are the amygdala, hippocampus, thalamus, hypothalamus, fornix, parahippocampal gyrus and cingulate gyrus.

Now let's have a basic idea of how various structures of the limbic system take part in our mental activities.

In the limbic system, an almond shaped structure located deep within the brain called the amygdala is the mind's emotion-coordinating system. From the amygdala emotional impulses go to the hypothalamus (the brain's Homeostasis center). Then the hypothalamus raises the blood pressure, heart rate and breathing, and puts the body into fight-or-flight mode based on the intensity of the emotional impulses. The amygdala also alerts the cerebral cortex (the brain's Intelligence center) which analyses the emotional situation and decides how much attention is required in that specific scenario. If the intensity of the emotional impulses is high enough then the higher nerve centers of the conscious brain become alert and strong conscious emotional sensation kicks in. Then the prefrontal cortex (the brain's decision-making center) in the frontal lobes plays its part by determining how to respond to the situation.

From an evolutionary perspective, one of the most fundamental role of the amygdala is to identify danger and trigger a sense of fear and alertness in the face of that danger. Such primitive fear response of the amygdala is necessary for self-preservation. Hence, the element of fear in our mental lives, is not actually an enemy. Rather, fear is wisdom in the face of danger, that we have acquired through millions of years of struggle in the wild.

Amygdala's neural connections to the rest of the brain put it in a unique position to rapidly respond to sensory input and influence physiological and behavioral responses, as well as to influence memory formation in the adjacent hippocampus, the memory indexer.

One of the most common characteristics of our limbic system is that, the stronger our amygdala responds to a situation filled with emotional potential, the more details of that situation is indexed by the hippocampus. The hippocampus is the brain's memory formation center, which is found deep within the brain, shaped like a sea horse. It consists of two horns that curve back from the amygdala. This little seahorse of your brain connects minute emotional senses like smell, sound or taste to memories and send the memories out to the appropriate part of the cerebral hemisphere for long-term storage.

So, here's the crazy part of this whole scenario. The more emotional you are in a situation, the more memories you'll have of that situation in the long run.

Also, in terms of gender difference, we have found through many studies that the hippocampus happens to be a bit larger in women than men. So, when a man cannot remember the details about the first date, it doesn't at all mean that he does not love his woman any more. It simply means that his brain circuits are unable to retain that specific memory.

Now comes the Thalamus. This little structure is endowed with one of the most significant features of your mental life. And it literally plays the most significant role when we talk about the human consciousness. Note that, I said, it plays the most significant role in consciousness, not one of the most. What could be so significant, one might wonder?

The thalamus literally modulates your very consciousness, by regulating the mental states of sleep and wakefulness. It plays a major role in regulating the level of your conscious awareness. Hence, damage to the thalamus can lead a person permanently to coma. It is also involved in relaying sensory and motor information to the cerebral cortex.

Right below the thalamus, resides the hypothalamus, which is the brain's Homeostasis center. It is responsible for various autonomic functions of the body, such as regulating body temperature, hunger, thirst, fatigue etc. It also controls daily cycles (circadian rhythms) in physiological states and behaviors such as sexual behavior. Another important function of the hypothalamus is to link the nervous system with the endocrine system via the pituitary gland.

The hypothalamus has a central role in neuroendocrine system of the body, most notably by its control of the anterior pituitary, which in turn regulates various endocrine glands and organs. Thus, it coordinates the rhythmic release of eight major hormones in the body.

Ultimately the profound regulatory influences of Hypothalamus over physiological and behavioral processes act as an essential feature in the path of survival.

Other few structures of the limbic system have their own responsibilities. The Fornix is involved in recalling a specific memory. Parahippocampal Gyrus is involved in memory formation and recalling visual scene. The Cingulate Gyrus participates in emotional anticipation. It also coordinates smells and sights, with pleasant memories of previously experienced emotions.

The sensory neurons of your entire nervous system that is spread throughout the body, receive crucial data from the external world, and based on that information, the neural circuits of various structures in the brain, construct all the elements of your mental lives. And these magnificent elements, such as emotions, ambitions, dreams, awareness, divinity, behaviors, personality etc. are collectively known as the human mind.

The Marvelous Interplay

Every single structure in the brain with a little help from the fellow structure, carries out its responsibilities quite amazingly. Turn off one structure, and you shall mute one element from the mental life. For example, turn off amygdala, and the element of fear would vanish from your mind. You shall only have a dim memory of that feeling, but would never be able to actually feel what it is like to be afraid. Turn off hippocampus and you shall never form any new memories from the time that you turn it off. You shall be stuck with the memories, acquired until the moment you turn it off.

Conclusion

Without the harmonious electrochemical activity of all the brain structures, the very thing which we call "mind", would suddenly disappear from the face of the earth. In fact, there is no such thing as the mind. There are only functional expressions of various neural circuits, that form various mental features. These features are intricately intertwined with each other - one neurologically dependent on the other. Thus, all these features collectively form our perception of a solid abstract body of mental activities, which we proudly venerate as "the human mind."

* * *

CHAPTER III

Consciousness

"If you think you understand Quantum Mechanics, you don't understand Quantum Mechanics" – these are the words said by the physicist Richard Feynman. And today the philosophers quite inadvertently put up the same argument for the explanation of Consciousness. From this argument rose the famous *Hard Problem of Consciousness.* Quantum Mechanics can indeed be extremely complex to grasp, but when we talk about Consciousness, with decades of rigorous studies on the human brain we have realized that actually, there is no other phenomenon in the entire universe that is simpler than the majestic phenomenon of Consciousness.

Introduction

Consciousness is simply the expression of our neurobiology. And at the core of it, it is nothing but a bunch of neurons firing relentlessly. And that's why when the neurons malfunction due to various neurological syndromes such as temporal lobe epilepsy, prosopagnosia, anosognosia, Cotard's syndrome and Capgras' delusion, the consciousness tends to malfunction as well. In turn that defective consciousness leads to an apparently defective reality.

By the harmful grace of the philosophers, the subject of Consciousness has remained a mystery to the general public despite all our achievements in Neuroscience. So in this chapter I shall attempt to clear the mystical air surrounding Consciousness.

"If you think you have a solution to the problem of consciousness, you haven't understood the problem." This age-old metaphysical and philosophical argument is strictly not true. If you are sufficiently clear-sighted enough, you can realize the problem itself was a matter of the past when we didn't have insight into the neurological basis of consciousness. And today it is common knowledge in Neuroscience that, all mesmerizing features of the Human Mind, including the glorious Human Consciousness, are born from the tiny specks of jelly inside your head. Perplexity is an undeniable part of a problem when we are pursuing to understand it. But once we reach that point where we actually discover the solution to the problem, everything starts to make sense and the term "perplexity" suddenly disappears from our perception.

Philosophy & Consciousness

In 1986 the American philosopher Thomas Nagel wrote

"Certain forms of perplexity - for example, about freedom, knowledge, and the meaning of life - seem to me to embody more insight than any of the supposed solutions to those problems".

That is absolutely true if we look at the problem from a philosopher's perspective. After all, the purpose of philosophy is not to seek the solution for a problem, but to entertain the problem and embrace it as an unavoidable part of our existence. And this is exactly where science comes into play. We scientists entertain a problem to solve it, not to accept it to be "insoluble".

I imagine that right now, this very minute, you are convinced that you are conscious - that you have your own inner experience of the world - that you are personally aware of things going on around you and of your own inner states and thoughts - that you are inhabiting your own private world of awareness - that there is something it is like to be you. This is what is meant by being conscious. Consciousness is our first-person view on the world. And this very view exists because there are neurons in your biology to produce this view. This is what we call subjective experience of a human mind.

For ages, the distinction between mind and body has been a major preoccupation of both eastern and western thoughts. And although these distinctions have generated an immense noise of endless debates among the philosophers, little of lasting value seems to have emerged. And above all, the general public seems to have remained in the same abyss of ignorance on the matter of mind-body link where they were some centuries ago. The discovery of the structure of DNA by James Watson and Francis Crick eventually led to the global awareness that at a molecular level the functioning of genes gives rise to what we call Life. This proves that nonscientists are willing to work to understand the key issues of life if scientists are willing to work at explaining them. I have therefore written this book as a simplified overview of how protoplasmic circuits of billions of neurons form the human mind, for the general reader from all walks of life.

The Hard Problem

In this chapter, I shall forge a fresh approach towards the problem of subjectivity, by treating it not as a philosophical, logical or conceptual issue, but rather as an empirical problem.

Here, the hard problem of consciousness is that conscious experience is private, subjective, and unsharable property belonging exclusively to a private self. Philosophers have been struggling for millennia to sustain the idea that this subjective, private experience of one human mind is unperceivable by another human mind. Or if we widen our perspective, then the argument of the philosophers would be, it is impossible to perceive the subjective private experience of one creature by another creature.

In the history of the study of this hard problem of consciousness, one of the most important question ever asked is "what is it like to be a bat?" First posed in 1950 it was made famous in a 1974 paper of that name by American philosopher Thomas Nagel. Nagel argued that understanding how mental states can be neurons firing inside the brain is a problem quite unlike understanding how water can be H20, or how genes can be DNA. *"Consciousness is what makes the mind-body problem really intractable,"* he said, and by consciousness he meant subjectivity. To make this clear he asked, *"What is it like to be a bat?"*

I shall get back to the solution of this so-called insoluble problem of subjectivity, but first let's realize what we mean by Consciousness.

Do you think your poodle is conscious? Or your cat? Or the birds chirping outside your window?

You know that you are conscious, but what about all those other living creatures! Are they conscious? Do the plants have consciousness?

If we observe close enough we can find that every living organism on planet earth has some sort of consciousness about their existence. At the very least, this primordial form of consciousness allows a species to react to environmental changes in order to ensure survival.

Basically, consciousness refers to a living organism's unique perspective of the world, that it lives in. And this very private point of view vividly differs from creature to creature. In natural circumstances a bat cannot know what it is like to be a tree. A tree cannot know what it is like to be an electric eel.

Every single living organism perceives the world around it with its own distinct mind. And our mind defines the value of things we see in the universe based on our perception of the reality.

Say, there is a book written by Tolstoy sitting right there on the table. To our unique human consciousness, the reality of the papers in the book, is infinitely different from the valuable literature that they possess. For the kind of consciousness possessed by the bug which eats those papers, literature is non-existent, yet for the Human Consciousness, literature has a greater value of truth than the papers themselves.

Such is the vast difference in conscious private perception of the world among creatures from various species. And even when we talk about us humans, each one of us possesses a unique subjective point of view of the reality. You can see the perceptual variance in the very thought experiment I concocted above. Due to my personal affinity towards the literature of Tolstoy, I chose him over other great thinkers, as the symbolic representation of all literature ever existed in mankind history. And this very personal appeal varies from person to person. You may like football, I like badminton. You may like Jazz, or some other kind of music, I like country classic.

Based on the neural responses of people's minds, their tastes and preferences may differ greatly and vividly.

Consciousness in All Life-form

Every living creature on this planet, has a conscious subjective perspective of the world. Even the plants may seem to us as standing indifferent to the human sufferings, but even they have their own unique mental universe. They have their own way of interacting with the environment. In this context, the first name that pops up in my mind is an Indian Biologist - J. C. Bose (not the Bose from Bose-Einstein Condensate). During my early school days, I learned of him and his fascinating experiment of detecting consciousness in the plants.

He showed that plants react like humans to various physical stimuli like pain, affection etc. Bose's perception was that all plants are endowed with a certain degree of individuality or subjectivity. By means of a Resonant Recorder and the Electric Probe designed by him, he was able to demonstrate that the collapse in the leaves of the Mimosa plant upon stimulation was accompanied by an electrical signal which traveled to the stem, and its passage through the stem (in both up and down directions) caused the other leaves to collapse. Similar experiments by other scientists confirmed the coupling of electrical oscillations and spontaneous leaf movements in Desmodium.

These experiments led Bose to the conclusion that just like animals, plants are in possession of a nervous system. The response of the whole plant to physical stimuli is a consequence of long range electrical signaling throughout the entire plant body. This is similar to the basic characteristic of neurobiology in animals.

And the electrical activity in the plant nervous system gives rise to what we may call the Plant Consciousness. In biology, such form of plant consciousness is termed as *"Tropism"*. Tropism refers to directed response of a plant due to environmental stimuli. There are various types of Tropism:

Phototropism - Response to light,

Heliotropism – Diurnal or seasonal response to the direction of the sun,

Gravitotropism – Response to gravity,

Thigmotropism – Response to touch,

Hydrotropism – Response to water,

Chemotropism – Response to chemical agents,

Thermotropism – Response to temperature,

Electrotropism – Response to an electric field.

There is also a weird kind of tropism in the characteristic of the parasitic creeper vine Monstera which grows along the ground in search of darkness caused by the shadow of a prospective host tree. This is called Skototropism. It is something like negative phototropism.

Similar to the neurobiology of animal anatomy, Tropism involves three distinct phases

1. Detection of the initial environmental signal by plant receptors,

2. Subsequent processing/transduction of the primary signal,

3. The consequent integrated physiological response.

Consciousness in plants manifests through electrochemical signals, just like in humans. There are actually two forms of electrical signals prevalent in plants: the action potentials (AP) and the slow wave variational potentials (VP). In contrast to the AP's, the VP's vary with the intensity of the stimulus and have delayed repolarizations. The ionic mechanism behind the transmission of AP's also differs significantly from VP's. Both AP's and VP's are involved in long distance signaling and can invoke a response distant from the local area of the applied stimulus. AP's have been implicated in very rudimentary forms of conscious activities such as trap/tentacle closure (for Dionaea, Drosera), regulation of leaf movements (Mimosa), increase in respiration and gas exchange (Zea), decrease in stem growth (Luffa) and induction of gene expression (Lycopersicon).

However, as plants are rooted to the same spot all through lifetime, the physiological response is more in terms of growth and development in contrast to animals who respond primarily by a variety of movements.

With the increasing complexity of the nervous system, a species climbs up the ladder to become more advanced in terms of consciousness. So far, the only species that has reached the top of the ladder is us, the Homo sapiens.

Qualia

Now the question is, can we the smartest species on earth diminish the difference of subjectivity between us and other creatures - or even among us humans? Can Carey Grant know, how it is like to be Albert Einstein! Can Stephen Hawking know how it is like to be Gandhi! Can Abraham Lincoln know how it is like to be Erwin Schrodinger! Can Sir Roger Penrose know how it is like to be Srinivasa Ramanujan!

Or how about the more generalized gender based differences in private mental lives! Can a man know how it is like to be cranky in PMS! Or can a woman know how it is like to have an erection!

In all these scenarios the apparent problem that rises in minds of the general public (including philosophers), is the problem of Consciousness, or what we Neuroscientists like to call the problem of Qualia. However most of us Neuroscientists dispute the very existence of the problem in this day and age. I would further add that, the problem did indeed exist in the human mind, when we didn't have detail understanding of the brain anatomy. And now that we do, the problem doesn't exist in the first place to those of us who study the human mind at a cellular level.

Now one might wonder, what the hell is Qualia? Qualia (singular – 'Quale') are simply the raw feels of conscious experience - the yellowness of red, or the sharpness of the tip of a pencil. With the term Qualia we can explain various elements of

Consciousness, much eloquently without any mysterious air around them.

Qualia give human conscious experience the particular character that it has. For instance, imagine a yellow circle. Here, your conscious experience has two qualia: a color quale, responsible for your sensation of yellowness, and a shape quale, responsible for the circle appearance of the imagined object.

In contrast to the ancient philosophical idea that qualia are private, subjective and unsharably exclusive features belonging to a private self, we Neuroscientists have concluded that the self, or the thing that leads to our illusion of unitary, persistent self, can be mapped anatomically to various intertwined neural circuits of various brain structures. And as it turns out, the distinction between mind and body, which has haunted philosophers for millennia, is only an illusion. Mind is born in the body, and the body drives the mind.

And Consciousness in only a major feature among various others of the Mind. Now the philosophers argue that, one creature's Conscious Mental Life is exclusively its own. It is very private. And it cannot be experienced by another creature.

But is it really true? We think not. Let me illustrate the situation with two thought experiments that shall ultimately diminish the subjectivity barrier among minds.

The Quale Exchange 1 (Thought Experiment)

Imagine, in a faraway future one of your descendants become a brilliant Neuroscientist with an excellent grip over the anatomical structures of every single region of the nervous system. His (could be Her as well) level of intelligence allows him to understand and even predict human behavior with excellent accuracy.

Now, for the sake of the experiment, let's say, he has two sensory disabilities in his body. One is that he doesn't have any sensory receptors on his lips. This means, he can't feel anything on his lips. However, let's presume the processing mechanisms in his brain correlated to the sense of touch on lips are still intact. The other disability shall be revealed in the context of the next experiment.

As a person who has no perception of any kind of touch on the lips whatsoever, your descendant gets seriously intrigued by the curious phenomenon that people call *Kiss*. So, he decides to study it. He begins his study on a female volunteer while she kisses her romantic partner. Let's call her Mrs. L. The brilliant Neuroscientist starts to study her brain, while she pleasures the blissful experience of kissing (even though kissing involves a lot of brain regions, for the sake of the experiment, we shall only consider the quale of touch in the experience).

He looks at various brain scans of her kissing sessions and points out heightened activity in different brain regions. And eventually he comes up with a detailed diagram of the entire sequence of neural events, from the receptors of the lips all the way into the correlated brain circuits. The next time, Mrs. L comes in for the study, the Neuroscientist shows her the diagram and says 'This is what's going on in your brains'. To which she replies, 'Sure that's what's going on inside my head, but I feel a kiss, where is that feeling in this diagram?'

'What is that?' he asks. 'That's part of the actual experience of the kiss which it seems I can never convey to you,' she says.

This is the alleged epistemological barrier which he confronts in trying to understand the woman's conscious experience. Here the scientific description of a brilliant Neuroscientist seems to be incomplete from a woman's perspective.

Now let's carry out another experiment to have a detail perception of the hard problem of consciousness.

The Quale Exchange 2

The other disability of the brilliant Neuroscientist is that he lacks cone receptors in his eyes to distinguish between different colors. He is color blind. However, let's presume the processing mechanisms for color in his brain are still intact.

As a person who has no perception of any color whatsoever, your descendant gets seriously intrigued by the curious phenomenon that people call *color*. So, he decides to study it. He begins his study on Mrs. L. The brilliant Neuroscientist starts to study the brain of Mrs. L, who can perceive all colors properly, as she verbally identifies the colors which are shown to her.

She looks at various objects and describes them as yellow or orange or blue, but those objects often all look like shades of grey to the color blind scientist. So, he points a spectrometer at the surface of one of the objects and it says that light with a wave-length of 590nm is emanating from the object's surface, but still he has no idea what color this might correspond to, or indeed what people mean when they say 'color'.

Fascinated by this, he studies the pigments of the eye and so on and eventually he comes up with a complete description of the laws of wavelength processing. His theory allows him to trace out the entire sequence of neural events starting from the receptors all the way into the brain until he monitors the neural activity that generates the word 'yellow'. Now, once he has completely understood the laws of color vision (or more strictly, the laws of wavelength processing), and he is able to predict correctly which color word Mrs. L will utter when he presents her with a certain light stimulus, he has no reason to doubt the completeness of his account.

One day he comes up with a complete diagram. He shows it to Mrs. L and says, 'This is what's going on in your brain.' To which she replies, 'Sure that's what's going on, but I see yellow, where is the yellow in this diagram?'

'What is that?' he asks. 'That's the part of the actual experience of the color which it seems I can never convey to you,' she says.

This is the alleged epistemological barrier which he confronts in trying to understand Mrs. L's conscious experience of color quale.

These two thought experiments eloquently allow us to put forward a clear definition of the Hard Problem of Qualia (or in a broader aspect Consciousness). They pinpoint the specific aspect of the perceiver's brain state that seems to make a brilliant Neuroscientist's scientific descriptions incomplete from the perceiver point of view.

The Quale Exchange 3

Now it's time for our third and final thought experiment. In this experiment, we'll take up a qualia problem similar to Nagel's 'what is it like to be a bat' problem, except here in our example we shall use a single qualia of night-vision in nocturnal creatures, instead of Nagel's version of the whole bat experience. In the Nagel version, it's the whole bat experience, the qualia produced by the bat's echolocation system along with everything else in its conscious mental life, which Nagel claims we cannot know. Most people, especially the philosophers would agree that you couldn't know what it is like to be a bat unless you are a bat. After all, the bat's mental life is so completely, utterly different. But in the final part of all our three experiments, we shall get to the solution of the whole qualia problem.

Now imagine, our brilliant future Neuroscientist is coming home from his lab after a whole day of work. Here, let's assume he is a perfectly healthy human being with no sensory disabilities. His way back home goes through a dark creepy forest. While driving his car, through the forest he sometimes feels nervous because of his vivid imaginations filling in for his lack of vision in the pitch black darkness on both sides of the road. Every thump or rustling in the bushes starts to make his heart go racing as he imagines unseen dangers all around him.

Then he starts thinking in his mind, 'The darkness of night, makes us humans so afraid. We start shivering with every single rush of wind against our skin, if we can't see our surroundings. But for the animals it is totally the opposite. Nighttime is their time for playing and working. The cat, the dog, the fox, none of them are afraid of the dark. They can easily maneuver in their darkness. The why can't I a genius Scientist of the 23rd Century do so?'

He already knows all about how the nocturnal animals can see in darkness. They have a tissue layer called the tapetum lucidum in the back of the eye that reflects light back through the retina, increasing the amount of light available to the photoreceptors, through blurring the initial image of the light on focus. The more light bounces around inside the animals' eyes, the better they see in dim light.

The tapetum lucidum contributes to the superior night vision of some animals. Many of these animals are nocturnal, especially carnivores that hunt their prey at night, while others are deep sea animals. We humans can't see well in darkness because we don't have that special layer of tissue in the back of our eye.

Analysis

Everything that I have said so far through my three thought experiments, is nothing new. They only show the epistemological barrier of Consciousness between an individual Man of Science and his Subject. These experiments clearly state the problem of why qualia are thought to be essentially private.

Apparently they also show that the problem of qualia may not actually be a scientific problem, because in all these experiments, the scientific description of the Neuroscientist is complete. It's just that his description is incomplete epistemologically because the experience of his subject is something he never will know. This is what philosophers have assumed for time immemorial - that there is a barrier which you simply cannot get across.

This so called barrier is only an illusion. An illusion that rises only when there is a process of translation in the middle. The language of nerve impulses, which neurons use to communicate among themselves is nothing like a spoken natural language such as English.

The problem in the first two experiments is that the subject can tell the Scientist about her qualia only by using an intermediate, spoken language, when she says, 'Yes but your diagram is still missing the actual feeling that I experience'. Here the experience itself is lost in the translation. And animals can't communicate with us humans. If they could they would probably say, 'You'll never know what it is like to actually have natural night vision.'

Solution to the Hard Problem of Qualia

In all these experiments, the scientist is only looking at a bunch of neurons and how they're responding to certain stimulus, such the color 'yellow' or the feeling of a 'kiss', but what the subject is calling the subjective sensation is supposed to be private forever.

However, here most of us Neuroscientists would argue that it's only private as long as the subject uses spoken language as an intermediary.

All that the future Neuroscientist needs to do, is to avoid that spoken language part and take a cable made of neurons from the subject's specific brain region correlated to the qualia (specific region of the somatosensory cortex for the kiss, or specific region of the visual cortex for the color or night vision) and connect it directly to the same area in his brain.

Then perhaps he'll have the exact same conscious experience as his subject.

In case of the first experiment, he'll feel the kiss after all (at least the touch of it), as the neural structures responsible for processing the sense of touch on the lips are intact in his brain. The connection has to bypass his lips, since he doesn't have any sensation of touch there, and go straight to the correlated neurons in his brain without an intermediate translation.

When the subject says 'kiss' or 'yellow' or 'night vision', it doesn't make any sense to the future Scientist, because all these terms are simply translation, and he doesn't understand touch language of the lips or color language or night vision language, because he never had the relevant physiology and training which would allow him to understand it.

But if he skips the translation all together and uses a cable of neurons, so that the nerve impulses themselves go directly to the responsible areas, then perhaps he'll say, 'O M G, I see what you mean.' The very possibility of this, demolishes the philosophers' ancient argument that there is a barrier which is insoluble.

We can apply the same procedure in various circumstances. We just need to be creative enough. We can make a woman feel how it is like to have an erection. We can make a man feel how a woman feels on her vagina, when a man penetrates her.

Notice that the Transcendence (transcending the private qualia barrier) cannot be done through any instruments that we may use to detect neural activity in the brain. This is because the instrument's output is a sort of translation of the events that it is actually detecting.

Thus, in principle we can experience the qualia of any creature. We just need to find out what that part of the brain is doing in a specific creature and then somehow graft it onto the relevant parts of our brain with all the associated connections. And Voila! We'd start experiencing that creature's qualia.

This way we can even solve Nagel's 'what it is like to be a bat' problem.

The point is we don't need to be a bat to experience its conscious experience of echolocation. Qualia themselves are just part of a creature's conscious experience of the Self. And by building a bridge of neurons we can experience other creatures' qualia, as an additional feature of our own mental life. Thus qualia are not the private property of a particular self, other selves can experience a creature's qualia as well.

And this way, if we share enough qualia with each other, then we might perhaps fuse two entirely different and unique human consciousness into one. It will require much more detail understanding of the brain anatomy for sure, but it is not inconceivable.

Malfunction in Consciousness

Consciousness is simply the brain's neural response to its surrounding environmental stimuli. Hence when the neural circuits malfunction, Consciousness tends to malfunction as well. I could cite many examples, but here are some of my favorites.

In various neurological syndromes, a person's conscious awareness of the reality gets vividly altered.

Let me tell you a story of a stroke patient with such an altered cognitive reality. Her condition was noticed by my friend and colleague V.S. Ramachandran, one of the most interesting neuroscientists of our times. He explains her condition as follows:

"Who was this rolling out of the bedroom in a wheelchair? Sam couldn't believe his eyes. His mother, Ellen, had just returned home the night before, having spent two weeks at the Kaiser Permanente hospital recuperating from a stroke. Mom had always been fastidious about her looks. Clothes and makeup were Martha Stewart perfect, with beautifully coiffed hair and fingernails painted in tasteful shades of pink or red. But today something was seriously wrong. The naturally curly hair on the left side of Ellen's head was uncombed, so that it stuck out in little nestlike clumps, whereas the rest of her hair was neatly styled. Her green shawl was

hanging entirely over her right shoulder and dragging on the floor. She had applied rather bright red lipstick to her upper right and lower right lips, leaving the rest of her mouth bare. Likewise, there was a trace of eyeliner and mascara on her right eye but the left eye was unadorned. The final touch was a spot of rouge on her right cheek—very carefully applied so as not to appear as if she were trying to hide her ill health but enough to demonstrate that she still cared about her looks. It was almost as though someone had used a wet towel to erase all the makeup on the left side of his mother's face!

"Good grief!" cried Sam. "What did you do to your makeup?"

Ellen raised her eyebrow in surprise. What was her son talking about? She had spent half an hour getting ready this morning and felt she looked as good as she possibly could, given the circumstances.

Ten minutes later, as they sat eating breakfast, Ellen ignored all the food on the left side of her plate, including the fresh–squeezed orange juice she so loved.

Sam raced for the phone and called me (Ramachandran), as one of the physicians who had spent time with his mother at the hospital. Sam and I had gotten to know one another while I had been seeing a stroke patient who shared a room with his mother. "It's all right," I said, "don't be alarmed. Your mother is suffering from a common neurological syndrome called hemi-neglect, a condition that often follows strokes in the right brain, especially in the right parietal lobe. Neglect patients are profoundly indifferent to objects and events in the left side of the world, sometimes including the left side of their own bodies."

"You mean she's blind on the left side?"

"No, not blind. She just doesn't pay attention to what's on her left. That's why we call it neglect."

The next day I was able to demonstrate this to Sam's satisfaction by doing a simple clinical test on Ellen. I sat directly in front of her and said, "Fixate steadily on my nose and try not to move your eyes." When her gaze was fixed, I held my index finger up near her face, just to the left of her nose, and wiggled it vigorously.

"Ellen, what do you see?"

"I see a finger wiggling," she replied.

"Okay," I said. "Keep your eyes fixed on the same spot on my nose." Then, very slowly and casually, I raised the same finger to the same position, just left of her nose. But this time I was careful not to move it abruptly. "Now what do you see?"

Ellen looked blank. Without having her attention drawn to the finger—via motion or other strong cues—she was oblivious. Sam began to understand the nature of his mother's problem, the important distinction between blindness and neglect. His mother would ignore him completely if he stood on her left side and did nothing. But if he jumped up and down and waved his arms, she would sometimes turn around and look.

For the same reason, Ellen fails to notice the left side of her face in a mirror, forgets to apply makeup on the left side of her face, and doesn't comb her hair or brush her teeth on that side. And, not surprisingly, she even ignores all the food on the left side of her plate. But when her son points to things in the neglected area, forcing her to pay attention, Ellen might say, "Ah, how nice. Fresh–squeezed orange juice!" or "How embarrassing. My lipstick is crooked and my hair unkempt."

Finally, I took a sheet of paper, put it in front of Ellen and asked her to draw a flower.

"What kind of flower?" she said.

"Any kind. Just an ordinary flower."

Again, Ellen paused, as if the task were difficult, and finally drew another circle. So far so good.

Then she painstakingly drew a series of little petals—it was a daisy—all scrunched on the right side of the flower."

Figure 3.1 The Picture that Ramachandran's patient drew

Such neglect is not blindness, rather it is simply a general indifference to objects and events on the left. Let me give you another similar yet a little different kind of neglect case history.

A schoolteacher suffered a stroke that paralyzed the left side of her body, but she insists that her left arm is not paralyzed. Once, when she was asked, whose arm was lying in the bed next to her, she explained that the limb belonged to her brother. Then, when she was asked to clap, she proceeded to make clapping movements with her right hand, as if clapping with an imaginary left hand near the midline, while her left hand kept lying completely paralyzed with no movement whatsoever.

This patient was in fact completely paralyzed on the left side of her body after a stroke that damaged the right hemisphere of her brain. And like this one there is a small subset of patients with right hemispheric damage who seem to be absolutely unaware of the fact that the entire left side of their body is paralyzed even though they are quite mentally lucid in all other aspects. In the year 1908 French neurologist Joseph François Babinski first observed this curious disorder in which the patient's tendency is to ignore or sometimes even to deny the fact that one's left arm or leg is paralyzed. Babinski termed this condition as "Anosognosia" that means "unaware of illness".

Neglect stories are very popular in the field of neurology. There is another kind of neglect or denial in which a person downright denies to be alive. It is another fascinating neurophychiatric syndrome called 'Cotard' or 'walking corpse' syndrome.

This syndrome was named after the French neurologist Jules Cotard. He described the condition as *'le délire de negation'* or *'the delirium of negation'*. It results in a feeling that one is either dead or immortal. In 1880 Jules Cotard reported the case of a 43 year old lady, Mademoiselle X who believed that she had "no brain, nerves, chest or entrails and was just skin and bone", that "neither

God nor the devil existed" and that "she was eternal and would live forever". The syndrome is described to have various degrees of severity, ranging from mild to severe. In a mild state, feelings of despair and self-loathing occur, whereas in the severe state the person with Cotard syndrome actually starts to deny the very existence of the self. In 2007 McKay and Cipolotti published a report on a 24 year old patient called LU. LU repeatedly thought that she was in heaven, even though she was actually in National Hospital, Queen Square, London and that she might have died from flu. The delusions diminished over a few days and were gone after a week.

Cotard delusion is usually associated with lesions in the parietal lobe as well as the prefrontal cortex. It can be treated with various antipsychotic, antidepressant and mood stabilizing drugs along with electroconvulsive therapy (ECT) and psychotherapy. The brain of an individual with Cotard syndrome, generates the conscious awareness of being dead. However such delusion does not hamper one's daily choirs and the person walks around and carries out daily activities just like a normal healthy person. The delusion goes away with treatment, but until it vanishes, it remains the only conscious reality to the person.

However, I must make something clear here. Individuals with these mental illnesses cannot be considered as "crazy", since they are completely lucid in all other daily activities. These mental conditions are simply emergency defense measures constructed by the unconscious to deal with sudden overwhelming bewilderments about one's body and the space around it.

There is another mind-boggling neurological phenomenon, called Capgras' syndrome, where the patient sees familiar and loved figures as impostors. This delusion is one of the rarest and most colorful syndromes in neurology. The patient, who is often mentally quite lucid, comes to regard close acquaintances, usually his parents, children, spouse or siblings, as impostors. One patient reported with absolute belief: "That man looks identical to my father but he really isn't my father. That woman who claims to be my mother? She's lying. She looks just like my mom but it isn't her." Many of the documented cases of Capgras' syndrome have occurred in association with traumatic brain injury. This implies that the syndrome has a neurological basis.

Capgras' delusion results from a disconnection between the face recognition region in the temporal lobe and the emotion center of the brain, i.e. amygdala. Face recognition pathways remain completely normal, so a person with Capgras' could identify everyone, but as the communication between the face recognition region and amygdala is selectively damaged he/she would not experience any emotions when looking at the faces of his/her beloved ones. In the case of the patient mentioned earlier, he doesn't feel a "warm glow" when looking at his beloved mother, so when he sees her he says to himself, "If this is my mother, why doesn't her presence make me feel like I'm with my mother?" So the only way he could make sense of it, is to assume that this woman merely resembles his Mom, but is actually an impostor.

Often the brain of a person with Capgras' delusion creates some really bizarre cognitive reality. In one of such recorded case histories, a patient was convinced that his stepfather was a robot, proceeded to decapitate him and opened his skull to look for microchips.

But why exactly the close relatives are perceived as imposters and not any other familiar face? This is because when a person encounters someone who is emotionally very close to him/her, such as a parent, spouse or sibling, he/she naturally expects an emotional glow, a warm fuzzy feeling. The absence of this glow in the most expected relationship is therefore surprising which is then rationalized through an absurd delusion. On the other hand, when a person sees someone familiar but not emotionally close, he/she doesn't expect a warm glow and consequently there is no need for the brain to generate a delusion to explain the lack of warm and fuzzy feeling.

Observing the medical histories of various neurological syndromes is like observing the fascinating nerve cells of the human brain in action, while they construct what we so proudly call the Human Consciousness. They remind us of the overwhelming aspects of human silliness. They remind us how such a simple natural response of the human Biology, is misinterpreted as the "last surviving mystery" of this planet.

Ego, Id and Superego

Talking about consciousness and the brain, it is hard to look at the brain, without wondering where Freud's ego and id are located. Here you must remember, Freud was a keen student of the anatomy of the brain, who wanted to develop a *scientific psychology – a new structural theory of mind.* But he couldn't, due to lack of technological advancement. That's why he wrote:

"all of our provisional ideas in psychology will presumably one day be based on an organic substructure."

Again in his 1920 essay *Beyond the Pleasure Principle* he wrote: *"The deficiencies in our description would probably vanish if we were already in a position to replace the psychological terms by physiological or chemical ones... "*

Freud developed his theory of mind based on the distinction between conscious and unconscious mental functions. In his theory he introduced three interacting psychic agencies - *ego, id* and *superego.* He perceived consciousness as the surface of the mental apparatus. Freud argued that, just as the bulk of an iceberg is submerged below the surface of the ocean, much of our mental function is submerged below the surface. The deeper a mental function lies below the surface, the less accessible it is to consciousness. And despite the lack of empirical evidence, psychoanalysis provided a way of digging down to the buried mental strata, the preconscious and the unconscious components of the personality.

However, with modern technological development in brain scan techniques, we are finding out that Freud's assumptions about the submerged unconscious mental functions were indeed true.

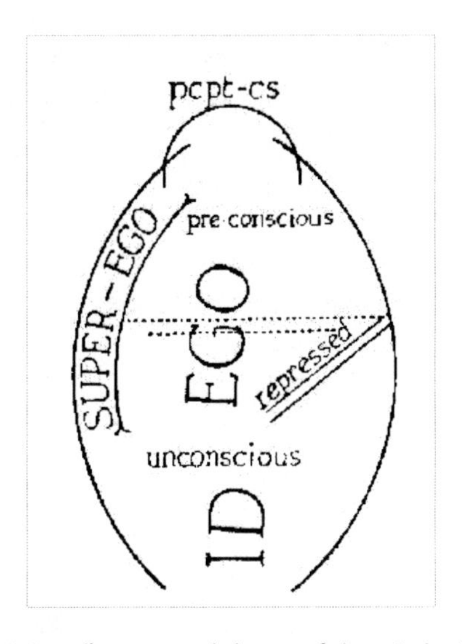

Figure 3.2 Freud's structural theory of the mind, with three psychic structures—the ego, the id, and the superego. The ego has a conscious component (perceptual consciousness, or pcpt.-cs.) that is in direct contact with the outside world, as well as a preconscious component, an aspect of unconscious processing that has ready access to consciousness. The ego's unconscious components act through repression and other defenses to inhibit the instinctual urges of the id, such as sexual and aggressive instincts. The ego also responds to the pressures of the superego, the largely unconscious carrier of moral values. The dotted lines indicate the divisions between those processes that are accessible to consciousness and those that are completely unconscious. (Source: New Introductory Lectures on Psychoanalysis, 1933)

For example, retention of implicit memory (driving a car, swimming etc.) is carried out by the brain quite unconsciously. Likewise, prolonged repression of emotions puts a strain on the unconscious mental operations, and eventually such repression leads to physiological symptoms. This condition is called *Conversion Disorder,* previously known as *Hysteria.*

Also, during life and death situations the basic instinctual pathways of the brain, which mostly remain submerged below the surface of your conscious neural activities in your usual state of mind, grab hold of your consciousness and express in the form of flight-or-flight response. For example, we have countless reports of parents lifting vehicles to rescue their children in life-threatening situations. In these incidents the primeval beast inside otherwise ordinary people lashes out to save their dear ones. And this immense fight-or-flight response results in an incredible strength which we call *hysterical strength.*

Freud's goal was not to develop a neuroanatomical structure of the mind, rather he wanted to give a functional map based on the differing mental operations. According to Freud's structural theory, the ego (the "I," or autobiographical self) is the executive agency, and it has both a conscious and an unconscious component. The conscious component is in direct contact with the external world through the sensory apparatus for sight, sound, and touch - it is concerned with perception, reasoning, the planning of action, and the experiencing of pleasure and pain.

The conscious component of the ego operates logically and is guided in its actions by the reality principle. The unconscious component on the other hand, is concerned with psychological defenses (repression, denial, sublimation), the mechanisms whereby the ego inhibits, channels, and redirects both the sexual and the aggressive instinctual drives of the id, the second psychic agency.

The id (the "it"), a term that Freud borrowed from Friedrich Nietszche, is totally unconscious. It is not governed by logic or by reality but by the primeval principle of seeking pleasure and avoiding pain. The id, according to Freud, represents the primitive mind of the infant and is the only mental structure present at birth. The superego, the third governor, is the unconscious moral agency, the embodiment of our aspirations.

Although we don't get much of a neuroanatomical idea of the mind from Freud's three psychic agencies, they have served as a major stimulant for many psychiatrists to dig deeper into the mind. Freud's perception of the mind was a kind of baby-steps towards the further understanding of the mind, and most specifically the human consciousness.

Conclusion

Consciousness is simple, if you are bold enough to accept it as simple. Genes come together to construct a magnificent life-form, while neurons come together to form our Illusion of Consciousness. And only by accepting its beauty in simplicity we can move on further in our path to discover the various unknown aspects of the human mental life.

* * *

CHAPTER IV

Emotions

Human mind is a symposium of various remarkable features. In fact, we call it a Mind, because of its features. Without the features there would be no mind. The mind is the collective expression of those features. In terms of Evolution, the most significant among those features would be Consciousness. However, from the perspective of a general human being - a non-scientist, the most valuable element of the human mental life, is Emotion - a tiny portion of our conscious mental world. We humans as a species crave for emotional stimulation. And in many cases, as it happens, we are actually slaves to our emotions.

Introduction

Until the moment when we Neuroscientists turned our attention on the study of the neurological correlates of emotions, there had been a lot of confusion among psychologists and philosophers concerning the best way to conceptualize emotions and interpret their role in life. In everyday human existence we conceive of an emotion - joy, rage, despair, grief - as a feeling - an inner state of our mental life.

Like any other cognitive feature, emotions evolved in the neural network of all creatures on earth in order to serve one evolutionary purpose - survival of the species. For example, fear and anxiety keep us alert in the face of danger. Love and attachment play a crucial role in promoting pair-bonding and mating. Hence, they ensure reproduction.

In the late 19th century, philosopher Wiliam James suggested that emotion is a function of sensory and motor areas of the cerebral cortex, and that the brain itself does not possess a special system devoted to emotional functions. This preliminary perception of the biological basis of emotion was eventually disproved, by further study of the brain mechanisms. However, James asked a very crucial question that triggered a chain reaction of neurological investigations - *'What is emotion?'* Since then, we have come a long way.

Contemporary neuroscientists have a vast arsenal of tools at our disposal for understanding intricate brain functions, from the level of anatomical systems to the level of molecules. Latest brain scan technology has given us fascinating insight into every single corner of the brain.

Over the past 100 years our understanding of the brain mechanisms their correlated cognitive features of the human mind, including emotion, has changed radically.

In the year 1937, Neuroanatomist James Papez proposed the first circuitry theory of emotion involving the hypothalamus, anterior thalamus, cingulate gyrus, and hippocampus. Several years later another Neuroscientist named Paul D. MacLean named the structures of Papez circuit, together with several additional brain regions – amygdala, septal nuclei, orbitofrontal cortex, portions of the basal ganglia – *The Limbic System.*

MacLean perceived the limbic system to be involved in the mediation of basic bodily functions required for the survival of the species. This means, the neurons of the limbic system react to environmental stimuli in a way that serves only one purpose - *Survival.* Thus all emotions serve that very purpose. However, from the point of view of a general human being, the emotions of life feel much more than just tools of survival.

Nevertheless, emotion is the reason why we the humans have survived this long on this planet as a species. On top of that, through the complex evolutionary process of reorganization of the neural network in our brain, we have ultimately become the Gods of this planet.

Essentially, the evolution of the brain has been one of the most significant events in the evolution of hominin life. It has been a 6 million years long mosaic process of size increase laced with episodes of reorganization of the cerebral cortex. The human brain is the most intricate, complicated and impressive organ ever to have evolved. The most obvious evolutionary change during human evolution has been the increase in size and complexity of the human brain.

Paleoneurology

The study of the evolution of human brain is an intriguing domain of scientific exploration. It's like talking to the ghosts of our extinct ancestors through their fossil remnants. This is what we call *Paleoneurology*. As a subfield of Paleoanthropology, Paleoneurology allows us to look into the details of brain structure of extinct species through close observation of endocasts (endocranial casts).

Paleoneurological study is at present the only direct line of evidence for cerebral evolution through time. The main object of study here is the endocast that is simply cast made from the inside of the cranial bone. But there is a very important factor to remember. It is that endocasts are not exactly casts of the brains. When alive, the brain is surrounded by three meningeal layers - the dura matter, arachnoid tissue and lastly the pia matter, a thin layer of tissue directly overlying the brain. Upon death, all these layers of tissues as well as the brain dissolve and leave a cranium that eventually fossilizes in time. However, technically endocasts are so far the only direct way of exploring the evolution of the human brain.

Endocasts reproduce a good deal of information about the brain, including its general shape and details of some of its associated blood vessels, cranial nerves and cranial sutures. Endocasts even reproduce information about the convolutions of the brain that were imprinted on the inner walls of the braincase during life.

These endocasts can tell us a lot about our true of biological history. Through the analysis of the fossil record, in concert with a comparative neuroanatomical analysis of closely related species, it shows that the hominid brain increased in size more than threefold over a period of approximately

2.5 million years. However, it has become increasingly clear that the human brain is not simply a large ape brain but this 3 lbs. cauliflower also developed important qualitative and quantitative changes. Among those extraordinary qualitative features, lies our amazing ability to control our emotional responses.

Emotional Control

In the everyday life of a human being, he or she faces various circumstances, that inadvertently trigger a cascade of emotional responses inside the brain. And the more modern region of the brain - the frontal lobes, keeps those emotions in check.

Now one might ask, why and how did such a fascinating quality of emotional-control developed in the human mind? To answer this question, we'd again have to look back at our ancestral history in the wild.

Social organization was imperative for our early hominin ancestors to survive in the harsh environment. Quite necessarily, natural selection pressured their brain to develop primitive social interaction skills through gesture and mimicry. And thereafter the brain developed a primordial form of basic intelligence. Such selective natural pressure gave our Australopithecine ancestors cortical capacity for social coordination.

Now, at that situation when our distant grandparents learnt about the advantages of living in groups, they had to learn the disadvantages as well. Any negative emotional outbreak in such groups would make a lot of noise, that would ultimately draw the attention of the predators. Naturally, again by the grace of Mother Nature selective pressure on the still primitive brain, our ancestors developed something extraordinary - the neural capacity to control emotions. And this very feature called for the dawn of advanced logical thinking in the early human mind.

As our ancestors gained control over their emotions by the augmentation of the frontal lobes, especially the prefrontal cortex, their social communities became more stable. With their early brain capacity of very primitive social communication the late Australopithecines built tools and later the Homos built shelters.

Such technological advancement preceded the arrival of an efficient language. And the threats of the wild surroundings demanded those early humans to become more advanced by developing further logical thinking. All the environmental demands triggered significant enlargement of the prefrontal and temporal cortices.

Here an important thing to notice is that, the prefrontal cortex enabled instrumental behavior, concentration, and emotional control as well as the integration of cognition and emotion necessary for decision-making and analytical thinking. Throughout a period of 6 million years there was a remarkable increase in cerebral cortical connections involved in cognition. Augmentation in the entire neocortex ultimately facilitated amazing brain functions to produce experiences of a coherent and meaningful mental universe with all its vivid and varied elements.

Reorganization of the whole neural network of the brain, especially in the cerebral cortex, led to various cognitive mechanisms that ultimately made us uniquely distinguishable from the rest of the animal kingdom. But, despite all our modern cerebral capacity of controlling emotions, it seems, that all our emotions have their own significant role in our survival. Our prefrontal cortex (PFC) only modulates those roles of various emotions. The PFC decides which emotional responses are to be exerted at what intensity through the conscious mind in which situation. Every emotion has its significance in your survival, even the most primitive ones, that most people feel ashamed of.

Rage, hate, anxiety, stress, love – all these are crucial emotional elements of your mental life.

And not accepting this biological reality has been a major cause of the global psychological distress. For ages many people throughout the world, inspired by metaphysical garbage, have been pursuing the golden goose of fake perfection and glory.

For example, the so-called spiritual preachers keep preaching their fairytales: *"we are all peaceful human souls, and all the negative elements of hatred, fear, lust, greed and rage are illusions that try to lead us away from the path of truth".*

The biological reality is, we humans are born with a hodge-podge of various brain circuits, that possess the seeds of peace, fear, love, hate, rage, pain, love, stress and faith. All these elements compose the emotional domain of our mental life. All these characters are ingrained in our limbic system, that keep our head straight in the path of survival. We humans can survive, only if, all these elements of our brain circuits function properly. Failure of any one element would mean extinction of the whole species.

And two of these elements have been immensely significant in our evolutionary history. These are – fear and love. Since our primitive days till today, fear has enabled us to be alert in the face of danger, and love has made us copulate and reproduce.

Now let's investigate them one after another at a cellular level.

Fear – The Evolutionary Wisdom

Stress, anxiety, insecurity, panic - all these emotional states that inflict tension upon our modern human society, are simply the neurological expression of the brain's fear response system. This response is born in the limbic system. And the particular limbic system structure that is the heart of all these dreadful sensations, is the amygdala.

The amygdala was part of the MacLean's limbic system theory. However, it did not stand out as an especially important limbic area until 1956 when Weiskrantz showed that the emotional components of a pathological condition named Kluver-Bucy syndrome, a symposium of behavioral consequences of temporal lobe damage, such as hyperphagia and hypersexuality, were due to the involvement of the amygdala.

But, still a detail knowledge about the brain's fear circuits were yet to emerge. For this scientists turned towards the experiment of classical conditioning (originally known as "conditional reflex") by the Russian physiologist Ivan Petrovich Pavlov (1849-1936). He discovered the concept of conditional reflex while examining the rates of salivations among dogs. Pavlov was fascinated to notice that when a bell (conditioned stimulus) was rang followed by presentation of food (unconditioned stimulus) to a dog in consecutive sequences, it would initially salivate when the food was presented. But eventually, the dog would come to associate the sound with the presentation of the food and salivate immediately upon the presentation of that conditioned stimulus of sound, even without the presentation of the unconditioned stimulus of food.

Pavlov's assistant Ivan Filippovitch Tolochinov who was with him the whole time during the experiment, presented the results at the Congress of Natural Science in Helsinki in the year 1903. Late the same year, at the 14th International Medical Congress in Madrid, Pavlov himself explained his findings in detail by reading his paper entitled "The Experimental Psychology and Psychopathology of Animals." His experiment earned him the Nobel Prize in Physiology/Medicine in the year 1904.

Further through the writings of John B. Watson and B. F Skinner, Pavlov's idea of conditioning as an automatic form of learning became a key concept in the then developing field of comparative psychology.

To have a basic idea of the brain's fear response circuits, Pavlov's concept of classical conditioning is exactly what we needed. In our experiments we called it *Pavlovian Fear Conditioning*.

In the late 1970s and early 80s, researchers began using a simple behavioral task of Pavlovian fear conditioning, to study fear circuits. This was a huge leap forward in the path of understanding the brain's fear response.

In Pavlovian fear conditioning, an emotionally neutral conditioned stimulus (CS), usually a tone, is presented to rodents in combination with an aversive unconditioned stimulus (US), often a footshock. As a result of such pairing, eventually the rodents form an associative memory between the CS and US.

After training of several such pairings, the conditioned stimulus acquires the capacity to elicit responses that typically occur in the presence of danger (as the unconditioned stimulus of footshock feels like danger to the rodents), such as defensive behavior (freeze or flee responses), autonomic nervous system responses (changes in blood pressure and heart rate), neuroendocrine responses (release of hormones from the pituitary and adrenal glands) etc.

These responses are neither learned nor voluntary. They are innate, evolutionarily typical responses to any kind of threat and are expressed automatically in the presence of appropriate stimulus.

Thus, fear conditioning allows new or learned threats to automatically activate very basic evolutionarily tuned instinctual responses to danger. Such fear conditioning happens in nature to all species all the time, including humans. It enables a species to learn about new threats. And our rodent model provides a valuable way to understand the neurobiology and behavioral psychology of this instinctual fear response.

Fear conditioning depends on the convergence of conditioned stimulus and unconditioned stimulus information in the amygdala. The CS-US association occurs in the amygdala via an intriguing synaptic process called *Long-Term Potentiation,* where the existing synaptic connections are strengthened through the release of more neurotransmitter in the synaptic cleft based on their recent activity.

Let's have a little more insight of this amazing synaptic strengthening mechanism. At the end of the 19th century, researchers recognized that the number of neurons in the adult brain - a hundred billion - did not increase with age. This fantastic revelation gave the neuroscientists good enough reason to believe that formation of new memories was not the result of the genesis of new neurons. With this realization came the need to explain, *how else could new memories be formed if not through the formation of new neurons?*

The father of Neuroscience, Santiago Ramón y Cajal proposed the idea that memories might be formed by strengthening the connections between existing neurons in the pursuit of improving the efficacy of their communication.

Cajal's idea echoed further in the celebrated theory of *Hebbian Learning* introduced by Donald Hebb, in his 1949 book The Organization of Behavior. Donald Olding Hebb (1904-1985) was a Canadian psychologist who made a crucial contribution in the neuropsychology of learning. In fact, he is often hailed as the father of neuropsychology.

"The Organization of Behavior" is considered to be Hebb's most significant contribution to the field of neuropsychology. As a combination of his years of work in brain surgery mixed with his study of human behavior, it finally brought together the two realms of human perception that for a long time could not be linked properly together.

Donald O. Hebb's The Organization of Behavior connected the fascinating domain of Neuroscience with the mysterious realm of Psychology.

In this book, he introduced the Hebbian theory, as a plausible explanation for the adaptation of neurons in the brain during learning process. It describes a basic mechanism for synaptic plasticity, where an increase in synaptic efficacy arises from the presynaptic neuron's repeated and persistent stimulation of the postsynaptic neuron.

In the book Hebb states:

"Let us assume then that the persistence or repetition of a reverberatory activity (or "trace") tends to induce lasting cellular changes that add to its stability. The assumption can be precisely stated as follows: When an axon of cell A is near enough to excite a cell B and, repeatedly or persistently takes part in firing it, some growth process or metabolic change takes place in one or both cells such that .A's efficiency, as one of the cells firing B, is increased."

Through various experiments and studies, Hebbian theory of learning through synaptic strengthening has been well established as an irrefutable fact of Neuroscience. But such a theory was farsighted for Hebb's time. In the late 19th and early 20th century neuroscientists and psychologists were not equipped with the neurophysiological techniques necessary for elucidating the biological underpinnings of learning in animals. These skills would not come until the later half of the 20th century, at about the same time as the discovery of long-term potentiation.

Long-Term Potentiation (LTP) was first discovered by the neurophysiologist Terje Lomo, through a series of neurophysiological experiments on anesthetized rabbits to explore the role of the hippocampus in short-term memory. In a 2003 paper, Lomo writes:

"In 1966, I had just begun independent work for the degree of Dr medicinae (PhD) in Per Andersen's laboratory in Oslo after an eighteen-month apprenticeship with him. Studying the effects of activating the perforant path to dentate granule cells in the hippocampus of anaesthetized rabbits, I observed that brief trains of stimuli resulted in increased efficiency of transmission at the perforant path-granule cell synapses that could last for hours. In 1968, Tim Bliss came to Per Andersen's laboratory to learn about the hippocampus and field potential recording for studies of possible memory mechanisms. The two of us then followed up my preliminary results from 1966 and did the experiments that resulted in a paper that is now properly considered to be the basic reference for the discovery of LTP."

After its original discovery in the rabbit hippocampus, neuroscientists have observed LTP in a variety of other neural structures, including cerebral cortex, cerebellum, amygdala and some others. Based upon all the neuroscientific evidence, we can say that it is very much plausible that LTP occurs in mammalian brains as well, including us humans. In the next chapter I shall further elaborate the cellular mechanism of LTP in detail.

Through this very process of Long-Term Potentiation the association of conditioned stimulus and unconditioned stimulus occurs in the amygdala, that in turns enables a species to learn about new threats, it encounters in the environment. It all happens through the strengthening of synaptic connections of neurons in the basolateral complex of the amygdala. And the end product of such synaptic strengthening is a primitive instinctual fear response to an apparently new perceived threat. Thus, the brain simply reacts to the modern threats of the human society, in a way it used to react during its primordial day of survival in the wild.

Production of the fear response to any kind of unpleasant stimulus (conditioned or unconditioned), depends on the central nucleus of the amygdala, which coordinates the output of defensive responses through downstream connections with response-specific brain centers. For example, fear responses like freezing and conditioned analgesia (suppression of pain upon exposure to an aversive stimulus) are modulated by the brain region called "periaqueductal gray", located around the cerebral aqueduct within the tegmentum of the midbrain. While on the other hand, another kind of fear response, like potentiated startle is mediated by the brain center called "reticularis pontis caudalis" or "caudal pontine reticular nucleus".

In the fear responses of the amygdala, hippocampus plays a crucial role as well. Over time the hippocampus forms memories of the entire context of the unpleasant stimulus (in case of animals, the chamber in which the pair of CS and US were presented, becomes the context, and as a result of this contextual fear, they exhibit fear responses when returned to the chamber). And proper functioning of the hippocampus is necessary for a new aversive stimulus to condition a creature for a fear response. For example, in case of amnesia in humans, lesions of the hippocampus made shortly after conditioning produce a severe and selective deficit of contextual fear, whereas those made a month or more after training have little or no effect on contextual fear. This happens because, the hippocampus is the brain's memory indexer that transforms short term memory into long term, by sending the newly attained memories out to various regions of the cerebral cortex for long-term storage. This conversion process is called memory consolidation. And once the memory of a context is consolidated, it becomes independent of the hippocampus.

Now let's get back to the amygdala. The central nucleus of the amygdala is the interface with motor systems of your body. Therefore, damage to the central nucleus interferes with the expression of various conditioned fear responses. On the other hand, damage to areas that the central nucleus projects to selectively interrupts the expression of individual responses. For example, damage to the lateral hypothalamus affects blood pressure but not freezing and conditioned analgesia responses, and damage to the periaqueductal gray interferes with freezing and conditioned analgesia, but not with blood pressure responses. Similarly, damage to the bed nucleus of the stria terminalis has no effect on either blood pressure or freezing responses, but disrupts the conditioned release of pituitary-adrenal stress hormones.

The central nucleus of the amygdala receives inputs from the basolateral complex, that consists of lateral, basal and accessory-basal nuclei of the amygdala. The lateral nuclei receives that majority of sensory information, which arrives directly from the temporal lobe structures, including the hippocampus and primary auditory cortex. The information is then processed by the basolateral complex and is sent as output to the central nucleus of the amygdala. Then, the central nucleus mediates the expression of conditioned fear responses based on the aversive stimulus. This is how most emotional arousal is formed in mammals.

The one and only purpose of the fear-response system is to avoid pain. Let's put the whole process simply in one sentence. Once a potentially unpleasant stimulus input reaches the lateral nucleus, the basolateral complex immediately starts processing the data and sends output to the central nucleus, which ultimately forms a suitable response.

Thus, the amygdala with the help of other limbic system structures contributes immensely to the survival of a species. Even in our modern human society, it enables a person to maneuver through the new and vividly aversive situations of daily life, by conditioning him or her to respond to them in a way that is most suitable for self-preservation.

So, as it turns out, the emotion, which most of the humans dislike the most, is one of the most crucial emotional responses of human existence. Without fear no complex life-form, such as ourselves, can exist whatsoever. *Fear is an evolutionary wisdom in the face of danger.*

Love – Nature's Tool to make Humans Copulate

Now that you have a detail insight into the neurobiology of humanity's most mistrusted emotion, let's shift our attention towards something completely different from fear.

Let's talk about mankind's most adored emotion - Love. However, love itself is not a single emotion, rather a blend of many. It is such an enchanting sensation, that it has been inspiring artists, scientists, philosophers and thinkers for ages. Albert Einstein said, *"any man who can drive safely while kissing a pretty girl is simply not giving the kiss the attention it deserves"*. Geniuses around the world came up with various creations under the spell of love. Schrodinger's Wave Equation, Hawking's Hawking Radiation, Tagore's songs, Rumi's poems, are just a few among the plethora of scientific and philosophical literature created under the enigmatic and warm influence of love. So, technically it is totally worth being crazy in love.

"Would you become a pilgrim on the road of love?

The first condition is that you make yourself humble as dust and ashes."

-Rumi

"I seem to have loved you in numberless forms, numberless time

In life after life, in age after age, forever."

-Rabindranath Tagore

However, despite all its fascinating influences over us humans, if we go deeper into its evolutionary bed of origin, we shall discover something rather extraordinary, or possibly something we might not like to see. The thing we call love, is actually, merely an illusion of the conscious mind. It is a byproduct of subtle sexual attraction. It is the expression of the mind's urge to copulate. And the closest real thing we got to love, is attachment, which grows over time in a romantic relationship, in the pursuit of serving another evolutionary purpose, i.e. nourishment of the progeny.

Underneath the wildly exaggerated term "love", there are two distinctly raw instinctual drives at play - sexual attraction (libido) and attachment. Sexual attraction, which in our modern human society, is mostly subconscious, enables a species to choose a healthy fertile mate to copulate with. And thereafter attachment programs a species to pay a great deal of time and effort in nourishing the offspring. That's all there is to Love.

However, in these neuropsychological phenomena of libido and attachment, the human brain goes through magnificent neurochemical alterations. And every time I look at them, these changes in the brain chemistry make me feel the awe in Mother Nature's inexplicable artistry.

In the beginning of our love lives, it is the beastly instinct of sexual attraction that drives us all. The butterflies in your stomach simply signal your mind that the person in front of you would make a fantastic mate to make babies with. Without this primeval drive, you won't ever fall for anyone in your entire lifetime. The very attraction you feel towards a person in a romantic way, is a mental manifestation of a subconscious desire to mate with that person. The euphoria, the madness and the bliss that come along with falling for someone, are a fascinating expression of your neurobiology triggered by a sexual stimulus in the environment.

However, when you open your eyes in the morning and with the first ray of sunshine have a glance at your dearly beloved, it seems as if you can see the radiance of a full moon in broad daylight. Then probably some of you wonder "how can this be science?" And you are right. It is not science. Science is the way to explain the "why" behind that feeling of yours. But, that very feeling is purely natural, regardless of whether Science explains it or not. You, me and all of us humans, feel it, because Mother Nature wants us to feel it that way. And that is the exuberance of naturalism. Every single neural mechanism inside our head, has been fine-tuned by Nature through millions of years, so that we do what is best for our survival as a species.

Hence, even though we naturalists call it purely sexual, it manifests itself in the mental lives of humans through various vividly colorful expressions. And in most cases, the general human mind does not even associate those expressions with sexuality. The romantic elements of our mental lives, though born from the neural mechanism of libido, are not limited to the expression of sexuality. They go much beyond the limited wild urge of having coitus. In the end, the general human mind perceives coitus, as an essential part of its love life, but definitely not the only part.

To the general public, the concept of love involves having an emotional bond with someone. But this very bond, is formed much later in a relationship. In the beginning, it is all about *living in a dream*. The very first few days when you actually start having symptoms of falling for someone special, are the days of heavenly bliss and unreasonable madness. This specific "madness" is one of the most rudimentary expressions of sexual attraction in the pursuit of building a romantic relationship. Along comes "euphoria". It feels like you have grown wings and you can fly around without a single care in the world. Everything starts to seem beautiful and better. The sun shines a bit brighter and the birds twitter a little louder and sweeter. As if you are stuck in an enchanting dream. You get butterflies in your stomach whenever the special person casts a blazing gaze upon you. And off course there is the thumping of heart and dilation of pupils. These are the basic symptoms of love, that shower a person with unimaginable bliss mixed with a little tinge of tension.

Like any other emotion, love is all about flooding of neurochemicals. Oxytocin, vasopressin, serotonin, dopamine, estrogen, testosterone - all these molecules along with the correlated brain circuits, form the apparently infinite emotional realm of love. However, they manifest in a man and a woman, in vividly different ways.

For men and women, the initial calculations about romance are totally unconscious, and they're quite distinctive. Men are chasers and women are choosers. It's our inheritance from the primitive ancestors who learned, over millions of years, how to propagate their genes. For example, men's common attraction towards an hourglass figure - large breasts, small waist, flat stomach, and full hips, is ingrained in the male neurobiology across all cultures. This shape tells the male brain that the woman is a young, healthy and fertile mate.

So, whenever a man ogles at a woman's breasts or hips, in most cases it doesn't mean he intends to sexually assault her. Rather it's a basic evolutionary instinct of the male brain that is embedded in the genetic blueprint. So, biologically speaking, it should actually be quite flattering than offending for a woman.

Darwin noted, males of all species are made for wooing females, and females typically choose among their suitors. If you take a closer look, you can observe such behavior all around you. The beautiful bird chirping outside your window. It's a mating call. That pretty little bird is trying to attract a potential mate, so that it can propagate its genes.

Why does the peacock have such beautiful feathers? It is to attract a healthy female. He as well is trying to propagate his genes.

Even we humans, are not much different from the rest of the animal kingdom when it comes to attracting potential mates. When women dress up for their night out at the club, they are doing so to look attractive. This is a subconscious evolutionary desire to attract as many potential mates as possible. In fact, the entire cosmetic industry thrives on people's desire to look attractive, which is in most cases, merely a subconscious urge to grab attention of healthy potential mates. Fashion, perfume, cosmetics, dietary supplements - all these cannot survive without the basic primordial human instinct to look sexually attractive.

While women tend to grab attention with their looks, men on the other hand, tend to attract as many potential females as possibly, by showing off their resources. When a man shows off with his fancy car, expensive gold watch and suit, or flexes his muscles and brags about how many credit cards he owns, he's doing so to make himself desirable by healthy women, in order to propagate his genes. It is all in the pursuit of reproduction.

It is the brain architecture of love, engineered by Mother Nature through the reproductive winners in evolution. Even the shapes, faces, smells, and ages of the mates we choose are influenced by patterns set millennia ago.

Falling in love is one of the most irrational behaviors or brain states imaginable for both men and women. The brain becomes "illogical" in the throes of new romance. If we could travel along a person's brain circuits as he or she is falling in love, we'd begin in an area deep at the center of the brain called the ventral tegmental area (VTA). We'd see the cells in this area rapidly producing dopamine.

Dopamine is a multifunctional neurotransmitter that plays various roles of immense significance in your mental as well as physiological lives. It is the predominant catecholamine neurotransmitter in the mammalian brain, where it controls a variety of functions including motor activity, cognition, arousal, motivation, positive reinforcement, food intake, and endocrine regulation.

Dopaminergic neurons or dopamine-producing neurons are few in number in the human brain - a total of around 400,000, compared to the hundred billion neurons of the entire brain. The cell bodies of these nerve cells are confined in groups to a few relatively small brain regions. However their axons project out to a vast domain of other brain regions, and they exert intense effects on their targets.

These dopaminergic cell groups were first mapped in 1964 by the Swedish physicians Annica Dahlström and Kjell Fuxe. The dopaminergic brain regions they identified were the substantia nigra, the ventral tegmental area, the posterior hypothalamus, the arcuate nucleus, the zona incerta, and the periventricular nucleus.

In humans, the projection of dopaminergic neurons from the substantia nigra connects with the dorsal striatum. This dopaminergic pathway is termed as the nigrostriatal pathway, that plays a significant role in the control of motor function and in learning new motor skills. These dopaminergic neurons of the substantia nigra are very much vulnerable to damage. Damage to this region of the brain leads to Parkinson's Disease.

Another dopaminergic neural region is the ventral tegmental area (VTA). This is the brain region that, upon getting triggered by a specific stimulus, takes the first step towards forming the emotion of love in the human mind. And the information based upon which these neurons respond to specific stimuli, is ingrained in the genetic blueprint of their nuclei.

VTA dopaminergic neurons project out to the prefrontal cortex via the mesocortical pathway and to the nucleus accumbens via the mesolimbic pathway. Together, these two pathways are collectively termed the mesocorticolimbic projection. VTA also sends dopaminergic projections to the amygdala, cingulate gyrus, hippocampus, and olfactory bulb. The entire mesocorticolimbic pathway plays the central role in reward and other aspects of motivation in the human mind.

The posterior hypothalamus has dopaminergic neurons that project to the spinal cord. The arcuate nucleus and the periventricular nucleus of the hypothalamus have dopaminergic neurons that form an important projection - the tuberoinfundibular pathway which goes to the pituitary gland, where it influences the secretion of the hormone prolactin. The zona incerta, grouped between the arcuate and periventricular nuclei, projects to several areas of the hypothalamus, and participates in the control of gonadotropin-releasing hormone, which is necessary to activate the development of the male and female reproductive systems, following puberty.

Among all the dopaminergic brain regions, VTA is the first station of the love circuit. In a relationship, dopamine released from this region of the brain, acts as the body's feel-good chemical for motivation and reward. Triggered by a visual, auditory or sensory stimulus, as the VTA neurons start producing more and more dopamine, and releasing into brain's pleasure center nucleus accumbens (NAc), the person starts to feel a pleasant buzz. The flood of dopamine stimulates the NAc in exerting the feeling of pleasure and arousal in the person's mental universe. However, you must remember that the NAc is involved in forming all kinds of pleasure sensations in the human mind, not just the ones connected to love and romance.

Everything that makes you feel pleasure, whether it is music, food, sex, movies or anything else, does so by triggering the nucleus accumbens. NAc is activated whenever environmental information of an individually favored stimulus reaches the VTA. Genetic information of the nuclei of the VTA neurons decides how the dopaminergic neurons shall react to the stimulus.

In the early phase of a romantic relationship, we'd see the dopamine being mixed with testosterone and vasopressin in a male brain, while in a female brain, it gets mixed with estrogen and oxytocin. The fusion of dopamine with these other hormones makes an addictive impact over the person, leaving both the male and female exhilarated and head over heels in love.

And the last stand of this mad love is the caudate nucleus (CN), the area for memorizing the look and identity of whoever is giving pleasure. Here we'd see all the minuscule details about the woman or the man being indelibly chiseled into the permanent memory. At this point your beloved one becomes literally unforgettable. Once the train of love has made these three stops at the VTA, NAc and CN, we'd see the brain's lust and love circuits merge together as they focus only on the beloved one.

The brain circuits for passionately being in love or the so-called infatuation-love share circuits with states of obsession, mania, intoxication, thirst, and hunger. Also, the circuits that are activated when we are in love match those of the drug addict desperately craving for the next fix.

The amygdala (fear-response system) and the prefrontal cortex (judgment and critical thinking system) are turned way down when the love circuits are running at their full potential. This is why we become literally blind to the shortcomings of our dearly beloved. The same thing happens when people take Ecstasy. So romantic love is a natural way of getting high. The classic symptoms of early love are also similar to the initial effects of drugs such as cocaine, heroin and morphine. Narcotics trigger the brain's reward circuit, causing effects similar to romance. Hence, the well-known phrase "addicted to love" is scientifically quite literal and accurate. Studies have shown that this early ecstatic stage of mad love lasts for around six to eight months. Once the euphoria of this mad stage is off the mind, the evolutionary mechanism of pair-bonding kicks in. In general terms, it is what we call attachment.

The lessons of relationship that our primordial ancestors learned are deeply encoded in the genetics of our neurobiological circuits of love. They are present from the moment we are born and activated at puberty by the cocktail of neurochemicals. It's an elegant synchronized system. At first our brain weighs a potential partner, and if the person fits our ancestral wish list, we get a spike in the release of sex chemicals that makes us dizzy with a rush of unavoidable infatuation. It's the first step down the primeval path of pair-bonding.

In the rose-colored world of pair-bonding the most influential arrows of cupid are Oxytocin and Vasopressin. These two are the most extreme chemicals of interest in all kinds of love and trust on planet earth. Attachment in any relationship is only possible by the grace of these neurochemicals. And attachment is exactly what keeps a relationship alive and healthy. So, all the so-called philosophical notion of *"love without attachment"* or *"detached love"* are biologically non-existent on this planet. We humans are biologically designed through millions of years of evolution to grow attachment. Love cannot survive without attachment. In fact, what we call "a mature romantic relationship", begins after the early euphoric stage, when a couple starts getting coupled together.

Our primitive ancestors learned various behavioral characteristics like jealousy, possessiveness and aggression to ensure the survival of their wild love life in the harsh environment. And all those behavioral responses eventually got engraved in our genetic blueprint. So, these are not the enemies in the path of a healthy relationship, rather they are the essential elements, without which no relationship can survive.

The evolutionary purpose behind the early stage of sexual attraction is to mate, but what about the later phase of romantic attachment? What is the biological purpose behind the human drive to grow a romantic bond?

To understand this, you must realize what love is all about. Evolutionarily speaking, love is all about procreation. And, only erotic passionate love making in the early euphoric phase of mad love, does not guarantee successful procreation. A lot more effort needs to be invested by the parents to actually ensure the survival of their progeny. For us humans, this can be up to twenty years. For this purpose, Mother Nature developed strategies beyond the *"one time fling"* approach to make mating partners collaborate until their progeny can survive on its own. Hence the neurological circuits of pair-bonding evolved. Along the way it led to the neuropsychological arrival of monogamy as a favored type of relationship. However, scientifically speaking, women tend to be more monogamous than men, whereas the tendency of men is to be polygamous and promiscuous. Among the hundreds of human cultures throughout the world, only one, the Thodas of South India, have officially endorsed polyandry (the practice of having more than one husband or male mate).

For a man, the optimal evolutionary strategy is to disseminate his genes as widely as possible, given his few minutes (or, alas, seconds) of investment in each encounter. It all makes simple evolutionary sense, since a woman invests a good deal of time and effort - a nine month long, risky, strenuous pregnancy, in each offspring. Naturally she has to be very discerning in her choice of sexual partners.

To put it in a nutshell, the early phase of primordial sexual attraction or libido is Mother Nature's evolutionary tool to coax a species into reproducing. And thereafter through the neurobiological processes of pair-bonding she nourishes the mental element of attachment, so that you can give full attention to the upbringing of your child. Off course we wish to feel that there is no Science in Love. We wish to believe that love is unselfish. But, despite all our unfounded assumptions, behind all the emotional exuberance of love, there is only one selfish biological desire – *the desire to reproduce.*

Conclusion

Fear and Love are the two emotional pillars of survival. If either one goes missing from our mental lives, we would go extinct. We are neurologically hardwired to avoid danger due to our brain's fear-response system, and on the other hand, we are neurologically hardwired to get sexually drawn to the opposite sex as per the ancestral wish-list in our genetic blueprint (homosexual neurobiology gets altered in the womb due to excessive exposure to opposite-sex hormone environment, and thus it develops more in line of the opposite sex giving the person homosexual orientation). All of this is to ensure that, we survive both as individuals and as a species. Every single mental task of human life either triggers or, is triggered by, some sort emotional circuits of the brain. Pleasure, rage, aggression, euphoria, peace – all are the products of synaptic transmission in various neural network of the brain. If even one of the pathways is severed, the entire emotional realm of our mental lives loses some of its essence. Emotions are what make life worth living, even though at the core of it all, is plain old evolutionary instinct of survival. All emotions serve that basic instinct. Having said so, it doesn't change the fact that we humans cherish every single emotional moment of our lives as if it is the last moment of our survival. So, forget survival and live a little.

* * *

CHAPTER V

Memory

Memory is one of the most remarkable aspects of our mental lives. It provides our mental lives with continuity. It gives us a coherent picture of the past that puts current experience in perspective. Without the binding force of memory, all experiences of the human mind would be splintered into as many absurd fragments as there are moments in life.

Introduction

Without the mental time travel provided by memory, we would have no awareness of our personal history, no way of remembering the joyous events of our life, or the agonies that made us who we are today. We are who we are because of what we learn and what we remember or recall.

Learning and recalling are thus the fundamental elements of our human lives that enable us to solve the problems we confront in everyday life. In the ultimate pursuit of survival, learning and recalling allow us to adapt to the environment. These higher functions of the brain enable an individual to build a personal history, as a unique creature. And several of such personal histories together construct cultural heritages that ultimately contribute to the development of the history of the entire human population. Without memory, we would be like newborn babies with no recollection of any kind of past. Thus, memory is essential not only for the continuity of individual identity, but also for the transmission of culture and for the evolution and continuity of societies over centuries.

Learning and Memory

One of the most intriguing feature of our nervous system is the astounding ability to adapt to the environment. We do so, by learning new skills and gathering information from the external world. Thus, over time, we improve our performance as a human being. This is possible because, the neurons underneath are not some sort of rigid mechanical structure that merely performs a single task over and over again. The beauty of the neurons lies in their plasticity.

Ramón y Cajal defined this fascinating neural feature as *"the property by virtue of which sustained functional changes occur in particular neuronal systems following the administration of appropriate environmental stimuli or the combination of different stimuli."* Plasticity is a compulsory biological process necessary in any organism in order to sustain life in a changing environment.

Neural plasticity allows all human experiences to get shaped both functionally and structurally in the nervous system. In the process it enables a human mind to grow throughout the lifetime.

To the general public the term "memory", refers to mainly the mental ability to recall the past. But, that is only one type of memory. Another type of memory is associated with learning new skills.

Hence, memory can be classified into two major types – one for learning and another for knowledge or recall. The first one refers to the storage of information associated with performing various reflexive or perceptual tasks. It is referred to as non-declarative or implicit memory because it is recalled unconsciously without any conscious awareness of the process, such as driving or typing. When we use implicit memory, we act automatically and we are not aware of being recalling memory traces.

Let me give you a simple example. While I am writing this book, I am simply thinking about the words in my mind, and the carefully crafted neural pathways associated with typing specific keys on the keyboard, beautifully trigger motor response in my fingers from which the words of my mind get transferred into the computer and show up on the digital page in extreme accuracy without me being consciously controlling my finger impression on the keyboard. I have had acquaintance with typing on a keyboard since my school days. Hence, over time my nervous system has developed the implicit memory of typing. You can imagine a similar situation while you are driving a car. In the beginning while you were learning to drive, you had to consciously focus on every single action of yours to make the car move on a specific direction without hitting a pole, but through practice over time, you have developed the implicit memory for driving. Thus, implicit memory fulfills the old saying *practice makes perfect.*

Implicit memory is a collection of memory functions and types of learned behaviors such as reflexive learning (sensitization, habituation), classical conditioning, trial and error learning, procedural memory (for skills and habits) and another form of evolutionary learning which I shall propose in this chapter as primal conditioning.

Now comes the second form of memory, called declarative or explicit memory. It is what you generally refer to when you use the term *memory.* Explicit or Declarative memory is recalled by a deliberate and conscious effort that concerns factual knowledge of persons, things, notions and places. Declarative memory can be further classified as episodic or autobiographical memory and semantic memory.

Semantic memory is mainly associated with your vast knowledge about the world. It is about having a general perspective of the big picture painted on the canvas of the world history. Today most of us neuroscientists and psychologists use the term semantic memory more broadly to refer to all kinds of general world knowledge, be it about concepts, beliefs or facts. This type of knowledge is independent of specific personal experiences. It is simply what you can refer to as common sense. It is more of a common knowledge that can be retrieved without reference to the circumstances in which it was originally acquired. In most cases, you may not even remember circumstances in which you acquired that knowledge. For example, the knowledge that lemons are shaped like golf-balls would be considered part of semantic memory, whereas knowledge about where you were the last time you tasted a lemon would be considered part of the memory which we call episodic memory.

Episodic memory allows us to remember personal events and subjective experiences. It gives us a sense of individuality by building a link between what we are today and what we have been in the past. It gives us the perspective on our own growth as an individual human being. Often, over time, autobiographical memory shades into semantic memory so that the experience of an event is remembered as the simple occurrence of such event.

Now let's go deeper into the cellular dimension of memory. Explicit memory systems involve distinct regions of the brain and is exclusively confined to those regions both at structural and functional level. On the other hand, implicit memory systems are integrated at various levels in the central nervous system including reflex pathways, striatum, cerebellum, amygdala and neocortex.

On top of that, the kinetics of learning a specific skill and consolidation of knowledge-based memories are quite different. Implicit memory, such as learning to ride a bike, takes time and a lot of attempts to build up, while explicit memory, such as memorizing this page, is more immediate and requires much less effort.

However, while explicit memory fades away quietly rapidly in the absence of recall and refreshing, implicit memory on the other hand is much more robust and often lasts for all your life even in the absence of further practice. For instance, once you learn swimming, you will never forget it, regardless of whether you practice it or not. And as for the explicit memory, say you memorize a phone number and after that never actually dial that number, then due to lack of recall the explicit memory of the phone number would eventually fade away and you'll no longer be able to recall it at all.

The True Beginning of Memory-Study

In our study of memory we owe a lot to one man. His name was Henry Molaison, known widely as H.M. He was not a scientist, but a patient suffering from partial seizures. In 1953, he underwent an experimental brain surgery for his seizures, only to emerge from it fundamentally and irreparably changed. He developed a syndrome which we call *anterograde amnesia*. He had lost the ability to form new memories.

For the next 55 years, each time he met a friend, each time he ate a meal, each time he walked in the woods, it was as if for the first time. And for those five decades, he was recognized as the most important patient in the history of brain science. As a participant in hundreds of studies, he helped neuroscientists understand the biology of learning (implicit memory) and memory (explicit memory). Molaison was to the science of memory what Anna O. (Bertha Pappenheim - the famous patient of the Austrian Physician and Sigmund Freud's mentor Joseph Breuer) was to psychoanalysis.

Henry Gustav Molaison at the age of nine, banged his head hard when he got hit by a bicycle. After that incident the boy developed severe seizures. At that time, scientists had no way to look inside his brain, like we can do today thanks to the advances in brain scan technologies.

Eighteen years after that bicycle accident, Molaison arrived at the office of William Beecher Scoville, a neurosurgeon at Hartford Hospital, as he was blacking out frequently, had devastating convulsions and could no longer repair motors to earn a living.

Scoville localized Molaison's epilepsy to his left and right medial temporal lobes (MTL) and suggested surgical resection of the MTL as a treatment. On August 25, 1953, at the age of 27, Molaison's bilateral medial temporal lobe resection included the removal of the hippocampal formation and adjacent structures, including most of the amygdaloid complex and entorhinal cortex. The remaining 2 cm of the hippocampi tissue appeared entirely nonfunctional because it had atrophied and because the entire entorhinal cortex, which forms the major sensory input to the hippocampus, was destroyed.

The surgery was successful in its primary goal of alleviating the seizures. But the procedure left Molaison radically changed. He developed severe anterograde amnesia, which means his implicit memory was intact, but he could not form new memories of any events after the surgery. His explicit memory was impaired, due to the surgical removal of the hippocampi. He even could learn new motor skills - he could form implicit memory, despite not being able to remember learning them.

Alarmed with the consequences of the surgery, Scoville consulted with a leading neurosurgeon in Montreal, Wilder Penfield of McGill University, who with neuropsychologist Brenda Milner, had reported on two other patients' memory deficits.

Soon after that Milner kept visiting Molaison in Hartford, to conduct various memory tests on him. This collaboration among three scientists and a graceful patient was about to alter conventional scientific understanding of learning and memory forever.

Milner stated about Molaison:

"He was a very gracious man, very patient, always willing to try these tasks I would give him, and yet every time I walked in the room, it was like we'd never met."

At the time, the conventional scientific understanding was that memory was widely distributed throughout the brain and not dependent on any particular neural region. Naturally when Scoville and Milner published their reports, many researchers attributed Molaison's deficits to other factors, like general trauma from his seizures or some unrecognized damage. It was hard for people to gobble the idea that it was all due to the excision of the hippocampi.

However, all of it began to change when Milner presented her study on Molaison, in which she demonstrated that a part of his memory was fully intact.

In her study she asked Molaison try to trace a line between two outlines of a five-point star, one inside the other, while watching his hand and the star in a mirror. At first it would be difficult for anyone to accurately perform the task. And every time Molaison performed the task, it struck him as an entirely new experience, since he had no memory of doing it before.

However, quite surprisingly over time with practice he became a master at it. After many of these trainings, at one point of time he said, *"Huh, this was easier than I thought it would be."* Milner's findings put an unforgettable impression on the scientific society, that basically altered many of our predominant ideas of the human mind.

From her study, we began to realize that there were at least two systems in the brain responsible for creating new memories. One known as explicit or declarative memory, that records new information such as names, faces and new experiences and stores them for future recall. This particular system of memory depends on the function of medial temporal areas of the brain, especially an organ that looks like a seahorse, called the hippocampus. Another system is what we today know as implicit memory. Retention of this type of memory is often unconscious and depends on various regions throughout the nervous system.

In daily life, you rely on implicit memory in the form of procedural memory, which allows you to remember how to tie your shoes or ride a bicycle without paying much of conscious attention to the task. Procedural memory is created through procedural learning or in simple terms, by repeating a complex activity over and over again until all of the relevant neural pathways get strengthened and work together to automatically produce the activity. Procedural learning is essential for the development of any motor skill or cognitive activity. It reorganizes the neural pathways and forms a long-term memory to guide an activity appropriately and accurately without conscious thought. By the grace of procedural memory people can pick up a guitar that they have not played in years and still remember how to strum it.

The study of Henry Molaison or H. M. (1926 – 2008) by Brenda Milner stands as one of the great milestones in the history of modern neuroscience. It widened the conventional perception and opened the way for the study of the two memory systems in the brain, explicit and implicit. Hence, it provided the very basis for the study of human memory and its disorders.

Memory Consolidation

Now the question is, how exactly memory is formed at a cellular level, be it implicit or explicit memory?

Basically, the formation of memory is predicated on the synaptic communication among neurons. When you try to learn a new skill or receive new information, it induces cellular and molecular changes that facilitate communication among neurons. These changes are fundamental for the storage of memory, be it implicit or explicit. Stronger the synaptic communication gets, more persistent your memory becomes. And the more emotional you are during a certain moment of life, the more neurotransmitters are exerted into the synaptic cleft, thus strengthening the synaptic communication even more for long-term recall of that memory.

Conscious memory of a new experience is initially dependent on information stored in the neural network of the hippocampus and the neocortex. All of this short-term memory information is highly dependent on the function of the hippocampus, until it is transformed into long-term memory. Without the hippocampus your brain can grab onto new information for only about several seconds. On the other hand in a healthy brain, with all regions in place, short-term memory information can remain stored up to about a year.

Based on the significance of this short-term memory in a person's life, it either fades away in time, or gets transformed into long-term memory. The fading away of short-term memory is simply known as *forgetfulness*, and its transformation into stable, long-lasting memory is called *memory consolidation.*

Our brain receives a whole lot of information every moment from the external world and only a little part of that information is actually useful. So the brain needs a mechanism to prevent itself from being burdened by negligible information. Hence, forgetfulness is as important as memory consolidation. To be consolidated, functional changes have to be followed by gene transcription and protein synthesis that produce permanent phenotypic changes in the neurons associated with structural rearrangements in neuronal networks. Thus, consolidation of memories is modulated by mRNA and protein synthesis inhibitors. Consolidation is not a high fidelity process. Stored memories gradually change and fade with time and only the most relevant and useful aspects are retained over time.

Initially the short-term memory is dependent on the activity of the hippocampus. And once the hippocampus guides the reorganization of the information stored in the neocortex, it eventually becomes independent of the hippocampus and gets stored for long-term recall. Until then, the dialogue between your hippocampus and neocortex is the cellular foundation of all the new information-storage that your brain has to do every day. After consolidation, the memories start to reside solely in the neural mesh of your neocortex.

So basically, consolidation is the process by which memories, initially dependent on the hippocampus, are reorganized as time passes. By this process, the hippocampus gradually becomes less important for storage and retrieval, and a more permanent memory develops in distributed regions of the neocortex. Here one thing to keep in mind, is that memory is not literally transferred from the hippocampus to the neocortex, because memory-information is encoded in the neocortex as well as in hippocampus at the time of learning those information. Gradual neural reorganization in the neocortex, beginning at the time of learning, establishes stable long-term memory by increasing the complexity, distribution, and connectivity among multiple cortical regions.

Cellular Mechanism of Memory Consolidation

The neural mechanism of memory consolidation involves two major cellular processes:

1. changes in the excitation-secretion coupling at the presynaptic level promoted by changes in channel conductances due to phosphorylation and Ca2+ influx,

2. Ca2+ influx at the postsynaptic level through NMDA glutamate receptors by Ca2+/calmodulin kinases, protein kinase C and tyrosine kinases promoting phosphorylation of neurotransmitter receptors and generation of retrograde messengers (such as nitric oxide and arachidonic acid) that reach the presynaptic terminal and increase neurotransmitter release in response to action potentials.

The activation of the molecules involved in these signaling pathways can last for minutes and thereby represent a sort of short-term *"molecular memory"*. Notably, all reactions mediated by phosphorylation typically have half-lives that depend on the kinetics of dephosphorylation by protein phosphatases. A very important role in the establishment of short-term memories is played by the balance between Ca2+/calmodulin-dependent protein kinase II (CaMKII), a key enzyme in synaptic plasticity at both pre and post-synaptic levels, and protein phosphatase 1 (PP1). Upon Ca2+ influx during training, CaMKII undergoes an autophosphorylation reaction that transforms it into a constitutively activated kinase. The *"switched-on"* CaMKII, however, is returned to the resting state by PP1 that thereby has an inhibitory effect on learning.

Thus, the antagonistic interactions between CaMKII and PP1 represent a push-pull system that plays a fundamental role during learning as well as in the delicate balance between maintaining and forgetting stored memories. These purely functional changes cannot survive for long times in the absence of a structural rearrangement of the neurons participating in the modulated synapse.

The sustained activation of the same pathways promotes memory consolidation by affecting the gene transcription and translation. Sustained stimulation leads to persistent activation of the protein kinase A (PKA) and MAP kinase Erk (MAPK) pathways. In turn, PKA phosphorylates and activates the transcriptional activator CREB1a, whereas MAPK phosphorylates and inactivates the transcriptional repressor CREB2.

The CREB family of transcription regulators is highly conserved across evolution and represents the major switch involved in the transformation of short-term memory into long-term memory. The CREB target genes, whose transcription is regulated during consolidation, include a set of immediate-early genes (such as C/EBP or zif268) that affect transcription of downstream genes. This results in changes, both increase and decreases, in the expression of an array of proteins involved in protein synthesis, axon growth, synaptic structure and function.

When synaptic strength has to be permanently potentiated, ribosomal proteins, neurotrophins, Ca2+-binding proteins, proteins involved in the exoendocytotic cycle of synaptic vesicles and neurotransmitter receptors become upregulated, whereas cell adhesion molecules that usually maintain synaptic stability become downregulated. These specific changes in protein expression favour growth of terminal axon branches and establishments of novel synaptic contacts.

This entire process of memory consolidation through synaptic strengthening is what I previously mentioned in the chapter Brain, as Long-Term Potentiation (LTP). Thus, through LTP memory gets consolidated, or in simple terms, transformed into long-lasting ones.

Opposite phenomenon occurs in case of long-term depression of synaptic strength, in which synaptic connections get weaker by decreased activity.

Memory, like all other magnificent elements of our mental lives, is merely a bunch a neurons communicating with each other in an accurate pattern. Sever the communication, and the memory it represents would be gone forever. Also, it takes time for memory to get stabilized in the hippocampus as well as the neocortex through the process of LTP.

In fact, LTP induced by an experience is inhibited by a novel experience administered soon (within 1 hour) after the first one, whereas an LTP established for more than 1 hour is immune to this reversal mechanism. Hence, the critical event in determining the retention of information consist in the stabilization of the potentiated hippocampal synapses in order to resist to LTP reversal upon the arrival of new information. Although hippocampus is fundamental to acquire new memories, it appears to be dispensable after the memory has been fully consolidated. For example, although the previously mentioned Henry Molaison was totally unable to build up new memories, he was still able to remember his life preceding the bilateral ablation of the hippocampi.

Permanent memories that are already consolidated are distributed among different cortical regions of the brain based on the various perceptual features associated with those memories so that, upon recall, the different components of a memory are bound together to reproduce the memory in its true integrity. This process appears to be time-dependent and hippocampus is still necessary to bind together the components of recent memories, whereas more remote explicit memories can be recalled independently of the hippocampus as the synaptic connections between various cortical regions associated with those memories already strengthened. So far we have understood that this conversion process of short-term to long-term memory occurs largely during sleep.

One of the fundamental features of memory is that it is formed in stages. Short-term memory lasts minutes, while long-term memory lasts many days or even longer. Behavioral experiments suggest that short-term memory grades naturally into long-term memory and, moreover, that it does so through repetition. Practice does make perfect.

Implicit Memory

In our everyday life we adapt to our environment by learning new things. Such learning alters our behavioral responses at a cellular level. And it does so, by forming implicit memory. However, unlike explicit memory, implicit memory is not confined to the exclusive neural twosome between the hippocampus and the neocortex. Implicit memories are distributed through various neural pathways in the entire nervous system. When you learn a new skill, the pathways associated with the implicit memory of that skill get indelibly strengthened. The memory associated with performing a certain task, is recalled unconsciously. We use implicit memory in various trained, reflexive motor or perceptual skills. Playing a guitar, driving a car, swimming – all of these once learned, are performed by recalling their associated implicit memory without conscious effort.

Explicit aspects of human memory, such as recall or recognition, reflect conscious recollection of the past. Implicit aspects of retention on the other hand, measure transfer from past experience on tasks that do not require conscious recollection of recent experiences for their performance. Thus we can say – *Explicit memory is recalling to the mind, and Implicit memory is retention without conscious recalling.* We study implicit memory in two elementary forms, which are *associative* and *non-associative learning.*

Associative Learning

In associative learning you basically learn that two stimuli are associated with each other. With associative learning you neurologically get conditioned quite unconsciously to respond in a specific way at a certain event. *Classical* and *Operant conditioning* are the two forms of associative learning. You have already read about classical conditioning in the last chapter. In Pavlov's experiment, the dog was classically conditioned to associate the conditioned stimulus of bell-ring with the unconditioned stimulus of food, and salivate even when only the bell was rang but no food was presented.

Classical Conditioning

Such classical conditioning occurs in your daily life all the time without being even aware of it. Let me give you a simple example. Outside the window of my study there is street. It is a suburban area, so mostly people either walk or ride their bikes on the street. Every time the sky gets clouded and it's about to rain, a person walks by my window shouting *"come rain, come"* in an attempt to asking for the rain to fall. I never see him, I only hear his voice. And oftentimes, his voice is followed by heavy rainfall. Over time, my brain got conditioned to associate the conditioned stimulus of his voice with the unconditioned stimulus of rain. Now, every time the person walks by my window shouting, I get this soothing sensation that it is about to rain. This is plain classical conditioning that elicits certain behavioral response without your conscious recalling of the process.

In classical conditioning, an animal learns to respond to a neutral stimulus in the same way it would respond to an effective stimulus. The most glaring example of such classical conditioning can be seen in the latest social phenomenon, known as *Islamophobia.* At a cellular level of the human mind, Islamophobia is not really a matter of social stigma, rather it is a natural biological fear response of the general human mind, conditioned through countless pairings between terrorist attacks (unconditioned stimulus) and their apparent association with Islam (conditioned stimulus). Hence, Islamophobia cannot be eradicated completely, unless that pairing is severed and thereafter the conditioned stimulus of Islam is paired with something optimistic such as the heartwarming works of the 13th century Persian *Muslim* poet Jalal ad-Din Muhammad Rumi.

Operant Conditioning

Another form of associative conditioning is operant conditioning. In operant conditioning (also referred to as trial-and-error learning), a person or animal learns that it gets a reward if it does something. So, a pigeon learns that it gets food if it pecks at a certain key, but not if it pecks at another. A rat learns that it can avoid getting an electric shock if it presses a bar at a certain time. Presumably what the animal learns is that one of its many behaviors (pecking or bar pressing) is followed by food. It is constitutional in animals to repeat behaviors that lead to positive reinforcement (something pleasant or the absence of something unpleasant) and avoid behaviors that lead to punishment or negative reinforcement.

Primal Conditioning

Classical and Operant conditioning are the most well-known form of conditioned learning. However, I believe they are not the only ones. I propose another type of conditioning which occurs through millions of years of evolution. I call it *Primal Conditioning.* It is the conditioning that our ancestors faced in the wild environment, for millions of years and eventually got imprinted into the genetic blueprint of their descendants through epigenetics. Such evolutionary conditioning got passed on from generation to generation. Nature conditioned our ancestors to certain stimuli over millions of years and the implicit memory of the conditioned responses to those stimuli are still intact in us, the Homo sapiens as an evolutionary heredity. Hence, every time we are exposed to those stimuli, we elicit certain conditioned responses quite unconsciously. Let me elucidate with a few of those unconscious retention of evolutionarily passed implicit memories.

For example, recent studies have found that single women who are looking for raising a family are more attracted towards men with facial hair, than clean-shaven men. Here what we see, is an unconscious retention of implicit memory, that goes back to our primeval days in the wild. In the primordial times, there was no such thing as shaving for men. They were all highly hairy mammals. And the existence of us the modern humans implies that our distant hairy grandfathers were quite successful in copulating with the females and raising families. Now let's look at this scenario from a conditioning perspective. In those days, the hairy appearance of the men acted as the conditioned stimulus, and along came the unconditioned stimulus of their protective behavior towards their women and offspring. This elicited in the women a conditioned response of emotional sensation of security while being with the men. This neurological process of conditioning went on from generation to generation. Thus the implicit memory of that conditioning got imprinted in the genetic blueprint of all the following females of the hominin line. As a result modern women have also received the implicit memory of this specific primal conditioning. Thus, even today full-bearded men are quite unconsciously perceived by women (especially those in the fertile phase of the menstrual cycle) as better fathers who could protect and invest in offspring. In short, due to primal conditioning of the female mind, facial hair of men strongly attributes to masculinity. And, masculine men achieve greater mating and reproductive success.

And it's not just women who are primally conditioned to be influenced by bearded men. All men are primally conditioned to ascribe dominance, self-confidence, courage, aggression and overall masculinity in bearded men. That's because the unconscious association between those personality traits and facial hair comes from the wild days.

Thus, because of primal conditioning, facial hair wildly influences people's judgments of men's socio-sexual attributes.

Non-Associative Learning

Now comes another form of implicit memory - non-associative learning, which occurs mostly through habituation and sensitization. Habituation is decrease in response to an effective stimulus when the stimulus is presented repeatedly. It is the simplest form of learning, through which an animal learns to recognize a stimulus that is harmless.

Habituation

When an animal perceives a sudden noise, it initially responds with several defensive changes in its autonomic nervous system, including dilation of the pupils and increased heart and respiratory rates. Now if the noise is repeated several times, the animal learns that the stimulus can safely be ignored. The animal's pupils no longer dilate and its heart rate no longer increases when the stimulus is presented. If the stimulus is removed for a period of time and then presented again, the animal will respond to it again.

As the neuropsychiatrist Eric Kandel likes to say:

"Habituation enables us to work effectively in an otherwise noisy environment. We become accustomed to the ticking of the clock in the study and to our own heartbeat, stomach movements, and other bodily sensations. These sensations then enter our awareness rarely and only under special circumstances. In this sense, habituation is learning to recognize recurrent stimuli that can safely be ignored."

Implications of Habituation in Life

Habituation allows us to ignore nonthreatening stimuli, so that we can give attention to stimuli that are really novel or associated with pleasure or danger. However, it is not just about ignoring harmless stimuli. Rather, repeated exposure to any stimulus, be it aversive or rewarding, eventually decreases the associated response.

That is why, sex becomes less and less pleasurable in a relationship over time. Your brain gets habituation to the sensual stimulation from your specific partner as you are exposed to it repeatedly.

It doesn't mean that the love is gone from the relationship. Love still exists beyond the barriers of time, in the form of attachment, which becomes independent of sexual intimacy after the euphoric stage of mad love.

However, after a long time of habituated sex, if your partner appears to you in a completely different avatar, it would again start sending your brain new novel stimulus that would eventually elicit a pleasure response by the flooding of dopamine in your nucleus accumbens. It would trigger you in ways that took place in the early phase of your relationship.

So, if you feel like the spark of sexual intimacy in your relationship is fading away, there is nothing to worry about. It is plain biology. All you need to do, to bring that erotic pleasure back, is to appear to your partner a little differently, such as changing hair color or style, wearing a different outfit, using new dirty words that you never used before etc.

Sensitization

Another form of non-associative learning is sensitization. Instead of making you ignore a repeated stimulus, it compels you to show an exaggerated response to almost any stimulus after having been subjected to a threatening stimulus. Just like habituation, sensitization as well occurs commonly in human life. For example, after hearing a gunshot, you would jump off your couch when you hear any kind of tone or sense a touch on the shoulder.

Cellular Mechanism of Habituation & Sensitization

Habituation and Sensitization have been extensively studied by Eric Kandel in sea snail *(Aplysia californica)*, that has a very simple central nervous system made by a few thousands neurons (about 15000-20000) as compared to the high complexity of mammalian brain (a hundred billion neurons in the human brain).

Aplysia is able to learn very peculiar behaviours. It learns to respond progressively more weakly to repeated innocuous stimuli (e.g. a light tactile stimulus), - habituation, and to reinforce the response to repeated noxious stimuli (e.g. a painful electrical shock), - sensitization. In both cases, the synaptic efficacy in the integration centre of a sensory-motor reflex is changed by experience, leading to an increased response of the reflex in the case of sensitization or in a reflex inhibition in the case of habituation. Both changes are integrated at the presynaptic level, mediated by changes in the Ca2+ influx in response to the action potential. In habituation, repeated stimulus activity leads to decreased dopamine release in the synaptic cleft, that results in reduction of synaptic strength or efficacy.

In habituation, Ca2+ influx is decreased into the sensory neuron terminal and the release of the neurotransmitter glutamate is accordingly decreased. This leads to the reduction in synaptic strength or efficacy which we call *synaptic depression.*

In sensitization, on the contrary, the activity of a facilitating serotonergic interneuron increases cyclic AMP concentration into the sensory neuron terminal, leading to PKA activation, phosphorylation of a potassium channel, lengthening of the depolarization evoked by the action potential and larger influx of Ca2+ increased glutamate release.

This leads to strengthening of the synaptic communication which we call *synaptic potentiation.* It is noteworthy that these two opposite forms of learning are associated with opposite changes in synaptic strength at the same integration centre of a somatic reflex arc.

Imitation Learning

Habituation and sensitization are the most generalized form of non-associative learning. But there are some other processes of learning that are much more complex. For example, through the process of imitation learning a child learns its mother-tongue. At the cellular level, imitation learning is a process conducted by a system called the mirror neuron system.

Mirror Neuron System – The Empathy Engine

Mirror neurons allow a person to observe and recreate the action of others. They are the reason why you start yawning or giggling merely by watching another person do the same. They play the key role in a child's brain, while it learns its mother tongue along with other cultural and sociological tactics. And as for adults, the mirror neuron enables you to learn new skills by the process of imitation. In the 1980s and 1990s, a few researchers Giacomo Rizzolatti, Giuseppe Di Pellegrino, Luciano Fadiga, Leonardo Fogassi, and Vittorio Gallese at the University of Parma, Italy discovered the mirror neurons in macaque monkey.

Mirror neurons were first found in various regions of the monkey brain - ventral premotor cortex (vPMC), inferior parietal lobe (IPL), primary motor cortex and dorsal premotor cortex (dPMC). Originally it was discovered, that the mirror neurons discharge both when the monkey does a particular action and when it observes another individual (monkey or human) doing a similar action. The name itself implies the significant feature of these fascinating nerve cells. This specific feature of *mirroring* or more specifically *imitating* has been evolutionarily crucial in shaping the modern human civilization.

The Mirror Neuron System (MNS) not only plays an important role in determining a human's capacity to imitate others' action, but also empathizing with them. MNS has vast impact over a person's social and behavioral skills throughout the lifetime. It allows us to be human and understand another human and even other species for that matter. When you see a person get beaten up in the park, you suddenly start to feel his agony. The same happens when you see a street dog getting hurt. Humans are biologically designed to truly understand another creature's pain, happiness and desires, as if it is our own pain, happiness and desires. Feeling others' emotions and imitation learning are the most influential features of the mirror neurons.

Imitation is a great mechanism of learning new skills in children and adult alike, although it is the most widely used form of learning during development, offering the acquisition of many skills without the time-consuming process of trial-and-error learning. Imitation is also central to the development of fundamental social skills such as reading facial and other body gestures and for understanding the goals, intentions and desires of other people. Many of us neuroscientists believe that there's a strong possibility that malfunction in the imitation learning mechanism or more specifically in the MNS may underlie various cognitive disorders, especially Autism.

During the developmental years of a child it learns language and different social skills. At this crucial stage of a person's lifetime, dysfunction in the mirror neurons might be one of the core deficits of socially isolating disorders such as Autism.

Mirror Neuron Deficit in Autism Spectrum Disorder

Autism Spectrum Disorder (ASD) is a pervasive developmental disorder characterized by impaired social interactions. In a study, scientists used functional magnetic resonance imaging (fMRI) to investigate neural activity of 10 high-functioning children with ASD and 10 normally developing children as they observed and imitated facial emotional expressions. Although both groups performed the tasks equally well, children with autism showed reduced mirror neuron activity, particularly in the area of the inferior frontal gyrus. Moreover, the degree of reduction in mirror neuron activity in the children with autism correlated with the severity of their symptoms. These results indicate that a healthy mirror neuron system is crucial for normal social development. If you have deficits in mirror neurons, you'd likely end up having social problems, as patients with autism do.

The imitation learning mechanism of MNS is responsible for enabling you to learn any new skills, like music, dance, mathematics etc. The more you observe and practice a specific skill, the MNS would fire more and the synaptic connections associated with the implicit memory of that specific skill would get strengthened even more. As a result, upon practice, you become better at that skill.

Prospective Memory

Everything you read so far is based on the most common classification of memory. However, memory is also classified based on the temporal direction. Memories, implicit or explicit, that are related to the past are called retrospective memory. It is basically the retention and recollection of past episodes. And memories connected to the future are called prospective memory. Let me elaborate.

After a change in his usual routine, an adoring father forgot to turn toward the daycare center and instead drove his usual route to work at the university. Several hours later, his infant son, who had been quietly asleep in the back seat, was dead.

Eight months after a hernia surgery, a patient complained of abdominal pain and nausea. A scan of his abdominal area revealed that a 16-cm clamp had been left from his previous surgery. Despite the best intentions of a surgical team of doctors and nurses, they had forgotten to remove the clamp.

The above errors represent failures of prospective memory. It is a form of memory that involves remembering to perform a planned action at some future point of time. Such memory tasks are highly significant in your everyday life. Without prospective memory you won't be able to plan any kind of future action. It ranges from relatively simple tasks to extremely significant ones. Examples of simple tasks include remembering to put the toothpaste cap back on, remembering to reply to an email or remembering to return a rented movie. Examples of highly important situations include a patient remembering to take medication or a pilot remembering to perform specific safety procedures during a flight.

Conclusion

Memory is the binding foam of our mental life. And all forms of memory from the simplest to the most complex ones, are encoded as plastic changes in synaptic connections sharing a common molecular communication. Memory is not just the cellular capability to store information, rather it is the very essence of our mental lives that gives us continuity.

You know who you are and what you are capable of, because of your own personal history and the experiences you gained throughout that history. And you may find your forgetfulness to be annoying, but it is necessary to lose some memories, to grasp that it is memory that fills up your lives, that gives you a meaningful perspective of the motion picture of life.

Life without memory is like a computer without hard disc. Memory is the coherence of life, that possesses all your emotions, and ambitions. Without it, your joyous as well as agonizing experiences of life won't have any significance to you whatsoever. Without memory we all would be frozen in time. We would be stuck without the ability to take even a single step forward.

* * *

CHAPTER VI

God

We humans are the gods of this planet. And we also have created Superior Gods than us, to have a sense of security. All this is because, through millions of years of evolution, Mother Nature's selective pressure on the neural network of the brain, has paved the way for the arrival of the advanced Human Consciousness. It is a construct of various cognitive processes. Among all the mental elements that these cognitive processes construct, the one element that has molded the entire human race like no other, is God. No other word in the world has created as much conflict, controversy, debate, love, peace, rage and violence as the word God. Private subjective perceptions of the meaning of the term God among believers and atheists have led to a never-ending battle between two people.

My primary goal in this chapter is to develop a Unified Scientific Theory of God, by treating the issue not as a philosophical, logical, or a conceptual issue, rather as an Empirical Issue of Qualia. I present an empirical idea on the ultimate sensation of God. I present a fresh neurobiological approach towards this apparently inexplicable sensation of the general human mind. In this chapter, through deeper investigation of the neurological correlates of the divine encounters with God, we shall reach the empirical conclusion that what we perceive as the Absolute Union with God or the Universe, is a fascinating composition of human qualia, that are constructed by billions of neurons firing relentlessly. It is what I have termed in this chapter as the Absolute Unitary Qualia. We shall discover that far from the age-old argument regarding God's existence, we should focus our attention on the human perception of God, in order to understand the mind in a better way. And above all, we don't need to solve the existential problem of God in order to realize that the God of human mind is ultimately a pure neurological sentiment, which does more good to human society than harm.

Introduction

In most cases, people argue over the term God, without having the perception of another person's own idea of the word. Hence, often people with an atheistic perspective of the world attribute the God of many religious individuals to be an angry, authoritarian and vengeful God who acts like a human being and lives in the clouds or in heaven. But the irony is, most religious individuals do not conceive God in an anthropomorphic or angry way. Rather, in their personal psychological domain of religious or spiritual beliefs, they conceive God in more abstract, spiritual and merciful way.

In these two cases, what we see is actually an apparently epistemological barrier in what I call *the qualia of God* in two people's psyche. Both their God-qualia consist of various mental elements. In the atheist mind, these elements are more inclined towards a more mechanistic and anthropomorphic perspective of the term God, which happens to be the Gods of the Scriptures. Hence, even though an atheist does not personally believe in the existence of God, he or she perceives the believers to have a belief in God in an anthropomorphic way.

On the other hand, in a believer mind, the God-qualia consist of more optimistic, positive, abstract mental elements. A mentally lucid believer (unlike the pathologically ill religious radicals) associates the term God with a more peaceful, merciful and loving sensation.

Thus lack of insight into each other's private qualia of God, results in a never-ending argument between two people with vastly different conceptions of the term God.

Meaning of God

Humanity has pondered over the meaning of God since its beginning. It is one of those cognitive features that came along with the advent of modern Human Consciousness. Consciousness is the symposium of distinctive elements of your mental life that you are aware of. It is a functional expression of your neurobiology. And at the core of it, there is nothing but a bunch of neurons sharing electrical information among each other. And when the neurons malfunction due to various neurological syndromes such as temporal lobe epilepsy, prosopagnosia, Cotard's syndrome, Capgras' delusion etc., the consciousness tends to malfunction as well. In turn, that defective consciousness leads to an apparently defective sense of reality.

However, even for a healthy human mind, reality is a construct of the neurons. An individual perceives the surrounding only as a simulation inside the brain. Now, from a more generalized perspective, this simulation of the actual reality of the external world, leads to all our conscious understanding and realization of ourselves and the world.

Now this neural simulation inside our head is influenced by various factors. And as far as biology is concerned, the most significant among them are - Evolutionary and Social Influence. Evolution has filled the neural simulation with all the mental elements that our ancestors learnt in their pursuit of survival, whether it is spirituality, sexuality or curiosity. I perceive these three mental elements to be the Pillars of Modern Human Consciousness.

Everything that is supernatural about the term God falls in the neurological domain of spirituality. Whether one conceives God in an absurd, non-physical, omniscient and eternal form or in a more human-like form, the majority of the human race is evolutionarily inclined to endorse some kind of belief in that idea.

Now, here comes the most crucial neurological correlate of the casual idea of God in believers. It is the neural mechanism of belief. As a whole, all religious individuals simply express some form of belief in God, or a Supernatural Force, that drives causality.

Now the question is – What is Belief?

As I have said in my book *Autobiography of God - Biopsy of A Cognitive Reality:*

A belief is your brain's self-maintenance mechanism. And the most important feature of a human belief is that it is very personal.

All human beings construct some sort of personal belief system. This belief system is not always religious or spiritual in nature. Every single idea that a human brain constructs during every walk of life based on available data, is simply a Hypothesis, or a Belief. It's a kind of myth, that your brain creates about the world. Your brain doesn't care whether that myth has even a shred of truth whatsoever. Each of all human brains creates its own myth about the universe.

And this myth inadvertently becomes a private element of the Neural Simulation of Reality. Now comes the Social Influence factor in the simulation. Often the myth your brain creates, embraces common characteristics from the society's predominant myths. Thus, through social interactions, religious myths spread from person to person at a cellular level in the form of explicit memory. On top of the evolutionary seeds of spirituality in the human mind, socially constructed myths get imposed further on the brain's religious belief circuits. However, religious belief upon a Supreme Divine Being itself is not rooted in any single patch of neurons, rather it is a psychological expression based upon various mental elements, such as emotions, memories, conjectures etc.

Through millions of years your brain has gone through magnificent neural reorganization. And this long reorganization process has led to the development of various amazing functions in different regions of the brain. Abstract, binary, holistic, logical and reductionist - all these functions have their distinct survival value. That's why all of these neural features still exist in the human brain. However, based on one's perceptual biases, some of these elements become more dominant over the others in an individual's mind. Hence, emerges all the conflicts among human understandings.

All these functions of the human brain help us process the enormous amount of information coming towards us from the external world every day. Relying on these functions, and their underlying strength of synaptic connections, the brain is able to construct a perceptual reality that suits one's comprehensibility. And the functions with the strongest synaptic connections, lead the brain to construct a cognitive bias. This bias allows the brain to further sustain its personal beliefs in its own mental life.

Hence the more you ponder over the term God, the more your brain constructs new synaptic connections in the line of your belief, that further strengthens your own personal qualia of God, be it religious, spiritual or atheistic.

Qualia of God

Now let me elaborate on the definition of the qualia of God. What does the phrase actually mean? Here I propose the phrase Qualia of God to refer to the private subjective experience or conception of God in people. With the use of this phrase we can diminish the confusions and vagueness that rise from the phrases like, Concept of God or idea of God. Qualia of God gives a more empirical approach towards solving the perceptual conflicts surrounding the term God.

Every single person, while talking about God, expresses his or her idea based on a private experience of that term. Thus one person's Qualia of God may differ vividly from another.

For example, an atheist might see God to be:

"the most unpleasant character in all fiction - jealous and proud of it; a petty, unjust, unforgiving control-freak; a vindictive, bloodthirsty ethnic cleanser; a misogynistic, homophobic, racist, infanticidal, genocidal, filicidal, pestilential, megalomaniacal, sadomasochistic, capriciously malevolent bully."

While on the other hand, most religious and spiritual individuals perceive the term God to be synonymous with words like hope, goodness, love, compassion, empathy and mercy. They feel a warm sensation running through their body while expressing their emotions associated with the term. Far from the atheist perspective that the God of religious individuals is an angry old man sitting in the sky while rolling his eyes over everything, the Qualia of God in the mental lives of most religious people are inclined towards positivity and optimism.

In a pursuit to explore this experiential domain of divinity, a very good friend of mine, Andrew Newberg of the Thomas Jefferson University and Hospital performed an informal study in which he asked people to draw what they think would be representative of God. And the results of almost three hundred drawings revealed something quite extraordinary. Around 33 percent of them drew a natural scenery such as a mountain with the sun in the sky, or perhaps a picture of the galaxy. Another third drew something abstract with circles, hearts, or swirling patterns.

From these results we get to see that only approximately 20 percent of those people actually conceived of God in some sort of humanized form. Most of them viewed God as a spiritual or abstract essence of the entire universe. Things get more interesting if we look at the remaining 14 percent. Quite interestingly, these pages were returned with nothing drawn on them. These pages belonged to the atheist individuals of the group along with some religious ones. For the atheists, they left it blank because they did not believe God existed, so there was nothing to draw. On the other hand, some religious people stated that God was undrawable and so they left it blank.

From this it is an easy deduction that the Qualia of God in general religious and spiritual people, are composed of natural feelings of love, compassion and an abstract wholeness.

But why did such Qualia evolve in the human psyche in the first place? To understand this we need to investigate their impact upon a person's mental life as a whole.

The Qualia of God have paramount potential to alter your body chemistry through mind-body substrates of neurobiology. Religious rituals such as prayer, tend to elicit beneficial health effects in people. Practice of prayer or meditation engages distinct neurochemical components of the human anatomy, that involve monoaminergic systems of arousal, reinforcement and reward, opioid systems of analgesia and pleasure, cholinergic systems of memory and other putative neurotransmitter systems (e.g. endogenous cannabinoids, nitric oxide, glutamate, gamma amino butyric acid, etc.) that modulate these effects.

Thus daily religious or spiritual rituals of prayer and meditation alter neural processes and their psychological effects of pain and analgesia. This gives religious and spiritual individuals more emotional control in times of utter distress and pain. On a functional level, such rituals trigger the corresponding neurobiological networks that are responsible for eliciting health effects by activation of various physiological, endocrinological and immunological processes in your body. Thus, such activities benefit the human body, reducing blood pressure and heart rate while improving the immune systems.

Also, through various studies we have found that, individuals who practice prayer or meditation as part of their religious or spiritual rituals, tend to have thicker and more active frontal lobes than those who don't. Frontal lobes are the brain's department of intellectual thinking and analysis of limbic emotions. Hence, healthier frontal lobes mean better grip over emotions in stressful situations of life.

So, as a whole, various clinical studies have demonstrated that religious and spiritual practices have profound beneficial effects on human psychology as coping strategies, and also in terms of feelings and mood.

Based on all this empirical evidence, I further deduce that – the Qualia of God act as the human mind's natural anti-depressant and mood-stabilizer.

For ages the myth-making mechanism of the human brain has been developing countless myths surrounding the Qualia of God, but at the very root of it all, lies significant survival value in both the short and long run.

Thus in general, if we leave out the atheistic fraction of world population that possesses no notable optimistic or positive Qualia of God, the majority of the human species possesses a beneficial Qualia of God enriched with blissful sentiments.

The Absolute Unitary Qualia

The majority of the world population senses divinity all around - in Nature and in the whole Universe. However, there are some rare individuals who actually claim to have an intense divine experience or encounter with God or some sort of divine being. This is what we call a transcendental state of consciousness.

In the history of humanity such transcendental experiences of a few individuals throughout the world, laid the foundation of all our religions. In this transcendental state of human consciousness, what a few rare individuals experience is the absolute blend of all human Qualia, that makes a person mentally one with the universe.

I hereby term this ultimate human Qualia as *Absolute Unitary Qualia*. It is a symposium of all the predominant qualia of a person's mind. In this Qualia, human mind dissolves into an ultimate state of Universal Consciousness. This state of consciousness has been worshipped with utmost reverence in various ancient cultures. When a religious person says, I want to become one with God¸ he or she refers to this Absolute Unitary Qualia. It is the ultimate limit of human consciousness, in which a person's very sense of the self disappears and the individual becomes truly one with the universe.

The Absolute Unitary Qualia can be referred to as what all religious cultures call being one with God. It is the ultimate representation of all human perceptions of God. It is the absolute perception of reality, that any human mind can ever have. The evolution of various cortical brain circuits endowed us with various brain states of consciousness, such as waking state and dream state. In our usual waking state of consciousness, most of us are just simple humans with a deep sensation of spirituality or religiosity. But the magic happens when we cross the boundary of that waking state and enter the mystical domain of transcendental consciousness. This is the very brain state of consciousness where we literally get immersed in God or the Universe. As if, our brain creates an air of mysticism around us, in which we lose all contact with our usual reality and get access to a supernatural reality, which feels more real than the reality we perceive when we are lucid.

There is nothing ordinary about this transcendental hyperspace or Absolute Unitary Qualia. In this altered state of consciousness, we experience what was often termed by the religious founders as revelation. It is a product of a hyperactive brain.

And there are many simple physical factors that can trigger such hyperactivity.

Transcendental experiences result from various heightened brain regions, along with some deactivated regions that are usually active in wakeful state of consciousness. In this highly excited state of consciousness, all the predominant intuitions, desires and drives find an ultimate absolution. A person truly transcends from his or her personal domain of limited perception, and dives into the infinity of the cosmos.

Everyday conscious awareness of a human being is only the tip of an iceberg, underneath which there is a realm of relatively uncharted apparently mysterious processes, which are likely to be way more complex than the usual waking state. Among those processes is the physiological mechanism for Transcendence. Over the course of human history, in this mystical realm of the human mind all the religious leaders have experienced what they called revelation, salvation or Nirvana.

To the person who experiences the Absolute Unitary Qualia, it is the zenith of experiencing God and divinity, that an individual could ever wish for. It is a mental state that enables a person to see, which he or she could never see with the usual state of consciousness. In such Qualia, every single predominant quale of an individual human mind bubbles to the surface and emerges as a part of one whole universal system. And when a human mind attains the Absolute Unitary Qualia, we can truly refer to that mind as the *Mind of God.*

The journey towards attaining Absolute Unitary Qualia is evoked by various physical events or stimuli such as, intense meditation, geomagnetic disturbance, natural DMT, psychedelic compounds such as LSD, cocaine, and amphetamines, magnetic stimulation, imposed sensory deprivation and brain lesion.

Here is a common expression of experiencing the Absolute Unitary Qualia:

"The experience took hold of me with such power that it seemed to go through my whole soul, so it seemed as if God was praying in, with, and for me."

Temporal lobe epileptic patients often experience similar sensation, during temporal lobe microseizures. History is filled with such individuals with temporal lobe epilepsy, such as Joan of Arc and Fyodor Dostoyevsky.

Dostoyevsky kept detailed record of all his epileptic microsezures. In his records he expresses:

"For several instants I experience a happiness that is impossible in an ordinary state, and of which other

people have no conception. I feel full harmony in myself and in the whole world, and the feeling is so strong and sweet that for a few seconds of such bliss one could give up ten years of life, perhaps all of life.

I felt that heaven descended to earth and swallowed me. I really attained god and was imbued with him. All of you healthy people don't even suspect what happiness is, that happiness that we epileptics experience for a second before an attack."

Along with the euphoria, temporal lobe epileptics often hear voices and have unusual visual disturbances and hallucinations, including prolonged blindness. They often report that God or God's angel has given them a mission to accomplish. This is what Joan of Arc experienced. She had encounters of Archangel Michael, Saint Margaret and Saint Catherine instructing her to support Charles VII and recover France from English domination. She suffered from tuberculosis, with a temporal lobe tuberculoma.

She was about 13 when she first heard voices in her head. The description of this first experience goes as follows:

"She had a voice from God to help her to know what to do. And on this first occasion she was very much afraid ... She heard the voice upon the right side and rarely heard it without accompanying brightness ... after she

heard this voice upon three occasions, she understood that it was the voice of an angel."

Eventually she engaged in a battle with the Burgundians even though she knew she was outnumbered by them and was finally captured. On this occasion she claimed that she had been misguided by her voices. The Burgundians handed her over to the English for a sum of money in 1431 and she stood trial. She was found guilty but signed a form of abjuration and was condemned to imprisonment. However, later she was sentenced to be burnt at the stake. She died exactly in that manner and it was well documented that her heart and parts of her intestines did not burn and were later collected and thrown into the River Seine.

A tiny brain lesion in Joan of Arc's temporal lobe triggered her auditory hallucinations, in which she heard voices of angels. But such altered state of consciousness is not always harmless. Some of you may argue that, Joan of Arc's brain lesion led to the freedom of France, which is true by the way. But that tiny lesion in her temporal lobe also was the cause of her excruciatingly painful death. In short, it was her neurological condition that led to her demise at the age of nineteen.

However, the most healthy and natural way to attain the Absolute Unitary Qualia is through intense prayer or meditation. Throughout the history of our species, many individuals have practiced and still practice various systems of meditation, prayers and other rituals. And the Supreme Purpose of all these rituals have always been to attain the Transcendental State of Consciousness, where Human, God and the whole Universe become one entity. All our religious leaders attained this state of consciousness through various ways - Buddha and all the Indian sages through intense meditation, Jesus through prayer and devotion, Moses and Muhammad through prayer, and so on.

Even today the Tibetan monks experience transcendence as a part of their daily ritual. And so did I, back in the days when there was no path open for me. Perhaps, that's why I was drawn to walk on the path of neuroscientific exploration, in the pursuit of finding absolution for myself.

However, for a general person to attain such a state, in the midst of a hectic life, requires practice. When an individual starts meditating, his or her brain goes through fascinating neurochemical changes. The very first stage is activation of the frontal lobes, as the person begins focusing attention on God or the Universe, or whichever point of focus he or she chooses. During the process, the activation of the brain's attention area -

prefrontal cortex (in frontal lobes), starts inhibiting various other brain activities, through the hippocampus.

Then the brain region that enables conscious spatial-temporal orientation, i.e. the parietal lobes start to become less and less active. And eventually, in case of intense religious or spiritual experiences, the person loses not only the sense of self, but also the sense of space and time, as the parietal lobes shut down almost completely. The limbic system is also highly involved in such mental activities. Hence, the activation of the limbic system leads to various emotional qualia in the Absolute Unitary Qualia. This leads to a feeling of intense euphoria and divine bliss.

When a person starts to enter the realm of transcendence, first comes the loss of all physical awareness, then the person ceases to identify the self with the mind and its individuality, but a shade of ego still remains, until the parietal lobes shut down almost completely. Then this little shade of ego or I also disappears. At that point whatever remains is the purest form of bliss one can ever experience.

Thus, in the Absolute Unitary Qualia the "I" of the human mind vanishes and unites with the universe. It is the true union with God or the Universe, or whatever you like to call it.

Conclusion

Current research in any field of Science has not yet reached the point where we could start exploring the existential question regarding God as a Supreme Entity driving causality in the universe. However, as modern Neuroscience progresses further and gets more advanced, we shall get to dive deeper into the physiological processes underneath the Qualia of God in human mind. What we have seen so far through our studies in Neurotheology, is that it is not God himself/herself/itself, rather it is people's perception of God that influences the human life. The Qualia of God impact all aspects of human life by altering the body chemistry at a cellular level. And something so biologically significant cannot be ignored simply as a conceptual issue, and definitely not as a delusion. All ancient wisdom were founded upon the Absolute Unitary Qualia. Hence, looking at them with a fresh neuroscientific approach shall give us new insight into the deepest fathoms of the human mind. By doing this, we may discover new clinical implementations of the human brain's natural element of spirituality. But for this, it is necessary for the future scientists interested in the field of Neurotheology to have a bit naïve approach towards the whole idea of God and religion beyond the conventional labels of religion and atheism. In conclusion, the purpose of Neurotheology shall be to ease human sufferings with a deeper understanding of the neurobiological substrates of spirituality and divinity.

* * *

CHAPTER VII

Free Will

You know what the most complicated feature of human nature is? It is the term *complication* itself. We are never satisfied with keeping things simple. We always tend to exaggerate even the simplest phenomenon of this planet. And one of those over-exaggerated phenomena of human history is *free will.* It all comes down to one specific question - *do we have free will?* And the only reason we are concerned with this question so damn much, is because it is epistemologically intertwined with the question of moral responsibility.

Introduction

Throughout the ages, philosophers have been consistently stuck with their conviction, that we must have free will, for our actions depend on us. They consistently argue that humans have control over their decisions, they are morally responsible for their actions, and that human actions are not pre-determined by destiny, by man-made gods, by logical necessity or by natural causal determinism.

Everything was working just fine with the philosophical and religious confirmation that free will does exist until scientific understanding of the human biology started to evolve and advance. And with the exploration of the cellular biology in the brain, we neuroscientists started to realize that all our actions are naturally determined by complex electrical activity within the brain. Thus, most neuroscientists arrive at the conclusion that, *free will does not exist.*

However, I won't make such an easy statement. In this chapter, I shall propose a fresh natural hypothesis on free will. I shall show that, the reality is something beyond the two reductionist labels of *existent* and *non-existent,* and that the question itself is not appropriate. First let me remind you something you already read in a previous chapter to elucidate the problem of free will. Billions of neurons together compose the illusory psychological expression, which we call *mind.* Free will is one of the most celebrated elements of the mind. Hence, it is a neural illusion as well.

The Right Question

Based on all the neural correlates of the mental life, I'd say that the *will* itself is not free from cellular mechanism. Like all other mental components, *will* rises from the protoplasmic activity within the brain. I believe the question – *do we have free will,* itself is not appropriate. We should mend our perspective a little, and start asking the question, *do we have the freedom of will, based on our experience?*

In the chapter on memory you learnt that, who we are depends on what we learn and remember. Every day you gather information about the world, and that information transforms you into a more learned version of yourself, than the previous day. If certain choice in life makes you suffer immensely with not much expected positive outcome, you get conditioned to avoid making the same choice again in the future. Thus, a choice that you made freely due to lack of experience and futuristic vision, is not going to be repeated once you learn about the downsides of that choice. Hence, given the same situation again in the future, the more experienced version of you will make a different choice, probably a better one, based on all your past experiences.

So, the conclusion is that in every walk of life, you do have the freedom to choose, but that freedom is based on the perception of the world and yourself which you have gained until that moment of life. Nature gives you the freedom to choose. But that choice always leads to one ultimate goal - *becoming a better version of yourself.* Freedom of will is born from the neurons. And that freedom allows you to sometimes make even the worst decisions ever in your life. And by making the worst decision, you simply learn what the better decision to make in future is.

Everything in your mental life proceeds in proper neurological order. If you could have sufficient insight into all the inner and outer parts of your mental life, along with remembrance and intelligence enough to consider all the circumstances and take them into account, you would be a true prophet and visualize the future in the present as in a mirror.

Articulation of Causality

This is similar to what Pierre-Simon Laplace visualized in his articulation of causal or scientific determinism as a superintelligent being. This being was referred to as *an intellect* by Laplace, and is generally known as *Laplace's Demon.* He wrote in his 1814 book *A Philosophical Essay on Probabilities (Essai Philosophique sur les Probabilites)*:

"We may regard the present state of the Universe as the effect of its past and the cause of its future. An intellect which at a certain moment would know all forces that set nature in motion, and all positions of all items of which nature is composed, if this intellect were also vast enough to submit these data to analysis, it would embrace in a single formula the movements of the greatest bodies of the Universe and those of the tiniest atom; for such an intellect nothing would be uncertain and the future just like the past would be present before its eyes."

Cellular Mechanism of Decision-Making

Given a situation, where you need to make a decision, the prefrontal cortex (PFC) of your brain, first analyzes all the options available to you while accessing the correlated implicit and explicit memory of your past experiences. Then in context of a set of needs and your personal history, the PFC potentiates the neural pathway for the execution of the most preferable among all the possible options.

Various regions of the prefrontal cortex are involved in distinctive cognitive and behavioral operations. And the regions that are specifically involved in various aspects of decision-making are - ventromedial prefrontal cortex (vmPFC), dorsolateral prefrontal cortex (dlPFC) and orbitofrontal cortex (OFC).

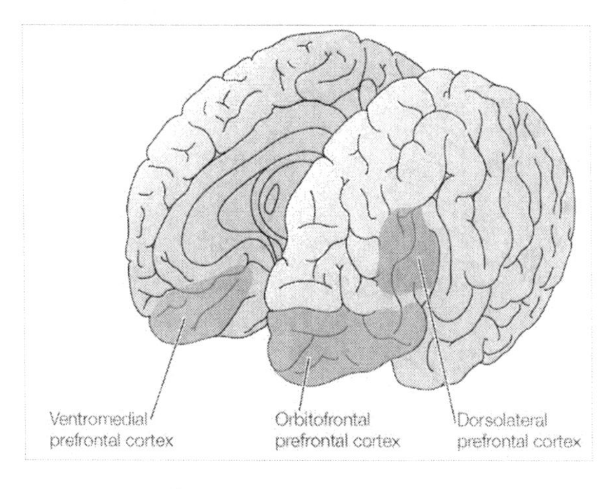

Figure 5.1 Prefrontal regions involved in decision-making

The vmPFC is crucial for your freedom of will to make a decision. Patients with bilateral lesions in the vmPFC develop severe impairment in personal and social decision-making, even though most other intellectual mental abilities remain intact. Damage to this region (especially in the right hemisphere) also leads to mental deficit in detecting irony, sarcasm and deception. People with damaged vmPFC are prone to be easily influenced by misleading advertising, due to their lack of doubt and skepticism.

The ventromedial prefrontal cortex is connected to and receives input from the ventral tegmental area, amygdala, the temporal lobe, the olfactory system, and the dorsomedial thalamus. It, in turn, sends signals to many different brain regions including, the temporal lobe, the amygdala, the lateral hypothalamus, the hippocampal formation, the cingulate cortex, and certain other regions of the prefrontal cortex. This huge network of connections affords the vmPFC the ability to receive and monitor large amounts of sensory data and thereafter influence your decision-making ability.

Emotions are a significant part of our mental lives. And vmPFC plays a key role in regulating emotions and inhibiting them if necessary, by influencing the limbic system, particularly the amygdala. The neural circuitry of the vmPFC is the birthplace of the moral nature of your behaviors and beliefs. Hence, malfunction in this region of your brain, cripples the very element of your mental morality. Without the healthy activity of the vmPFC individuals endorse actions of self-preservation that often break moral values associated with the term *human*, and inflict harm to others. And the most glaring instance of such vmPFC deficit can be seen in the phenomenon of religious terrorism, or more commonly *jihad.*

On the other hand, another crucial brain region involved in decision-making is dorsolateral prefrontal cortex (dlPFC). It manages various cognitive functions of the mind, such as working memory, cognitive flexibility, and planning. And when it comes to decision-making, dlPFC carries out all the risky stuff. When given several options to choose from, the dlPFC evokes the mental preference towards the most reasonable option while suppressing temptation in order to maximize personal gain. It gives birth to the ability of self-control in a certain situation for a better outcome. Unlike the vmPFC which makes decisions based upon moral values, healthy activity in the dlPFC facilitates self-preservation. In particular it plays three distinct roles in your mental life:

1. provides cognitive control to override predominant social-emotional responses elicited by the dilemmas,
2. facilitates abstract reasoning, such as, cost-benefit analyses, and
3. generates self-centered and other-aversive emotions such as, anger, frustration, or moral disgust.

Also, quite fascinatingly, several studies have shown that, increased activity of the dlPFC is often associated with psychopathic traits.

Thus, both the vmPFC and dlPFC are intertwined at a functional level along with another prefrontal region called the orbitofrontal cortex, when it comes to make a decision. Increased activity of the dlPFC without the correlated functions of the vmPFC leads to apparently inhuman and antisocial traits of psychopathy. On the other hand, without the healthy activity of the dlPFC, you would probably start to show altruistic attitude. Thus, in order to sustain a healthy life, dlPFC and vmPFC keep each other in check. In fact, when the both are working in proper harmony, you'd have excellent self-control as well as fantastically effective moral values, that would ultimately enrich your mental life. In simple terms, moral values are encoded in your vmPFC, and effective selfishness is encoded in your dlPFC. And you need both, in order to survive.

Orbitofrontal cortex also plays a significant role in decision-making. It enables you to anticipate the possible reward or punishment in a certain situation. It signals the expected rewards/punishments of an action given the particular details of a situation. In doing this, the brain is capable of comparing the expected reward/punishment with the actual delivery of reward/punishment, thus, making the OFC critical for adaptive learning. Thus, OFC plays the part in your mental life, where it analyses the potential emotional outcome of a certain decision in a specific situation.

This way, every single decision of your life is predicated on the healthy functioning of the prefrontal cortex. Even a slight malfunction in a tiny chunk of neuron anywhere in the PFC would lead to the mental deficit in your logical decision-making.

Formation of Preference

In your daily life, you make dozens of choices between an alternative with higher overall value and a more tempting but ultimately inferior option. In such scenario, formation of preference involves both cognitive and emotional brain circuits. A host of factors influence the development of preference, including physical features of the options, like the nature of their possible outcomes, such as valence (positive, negative), salience (intensity, magnitude), probability (degree of certainty), and timing (delay), relative values and number of options to select from, previous experience with these options and their outcomes, and external and internal context in which the decisions are made (social, mental state). Each of these factors intricately influence the decision you ultimately make.

Mental Stability and Free Will

For a person with a lucid state of consciousness, the PFC weighs all the available options in minute detail. But when the consciousness itself tends to malfunction due to certain neurological condition, the brain cannot make the best possible decision, as the neural pathways themselves are not stable enough. And it doesn't always have to be a neurological condition at the root of a defective consciousness. Humans face mental instability quite frequently in these modern days. Rage, depression and anxiety have become quite common in modern lifestyle. And these mental states inhibit the PFC to make the right decision. Let me give you a simple example.

Imagine yourself having a fight with your romantic partner. The tension of the situation makes your limbic system run at full throttle and you become flooded with stress hormones like cortisol and adrenalin. The high levels of these chemicals suddenly make you so damn angry, that you burst out in front of your partner saying, *"I wish you die, so that I can have some peace in my life"*. Given the stress of the situation through highly active limbic system, your PFC loses its *freedom* to take the right decision and you burst out with foul language in front of your partner, that may ruin your relationship. In simple terms due to your mental instability, you lost your *free will* to make the right decision.

But when the conversation is over, and you relax for a while, your stress hormone levels come down to normal, and you regain your usual cheerful state of mind. Immediately, your PFC starts analyzing the explosive conversation you had with your partner. Healthy activity of the entire frontal lobes, especially the PFC suddenly overwhelms you with a feeling of guilt. Your brain makes you realize, that you have done something devilish. As a result, now you find yourself making the willful decision of apologizing to your partner and making up to him or her, no matter how much effort it takes, because your PFC comes up the solution that it is the healthiest thing to do for your personal life.

From this you can see, that what you call *free will* is something that is not consistent. It changes based on your mental health. Mental instability or illness, truly cripples your *free will.* And the healthier your frontal lobes are, the better you can take good decisions. And the most effective way to keep your frontal lobes healthy is to practice some kind of meditation, as I have discussed in the last chapter.

The PMDD Conundrum (Thought Experiment)

Mental health influences your decision-making ability regardless of your experiences. However, experience is what makes you better at making the right decision in a given situation. Here the term *right* only refers to your subjective perspective of the choice. To explain how experience alters one's *free will,* let's carry out a little thought experiment in the context of two different scenarios.

Context 1

Imagine yourself to be layman with no notable awareness of biology. You start dating a woman planning a long-term relationship. Every month right before her period starts, she gets extremely cranky like all menstruating women. However, she tells you that she has a rare medical condition called Pre-Menstrual Dysphoric Disorder or PMDD, which is the extreme form of PMS. You have already heard all about women getting agitated during their PMS, so you think that it is just the same. However, in time things get real stressful. Every month during those days, she would turn into a completely different person. And one time, she gets so agitated that she bursts out – *you are the worst decision of my life.* Naturally, it feels beyond acceptable to you. You are a human after all.

You cannot ignore such behavior any more. You start perceiving every single insulting word from her mouth to be true. It seems that it is what she really feels. You start getting upset beyond tolerance. And after several of such intense insults from her during those days, you realize that you don't deserve such nonsense. You have no practical idea of what PMDD really is, so you perceive the insulting and violent behavior of your woman to be extremely inappropriate. So, one day you simply go ahead and tell her – *you are nothing but a psychopathic maniac who just likes to hurt people. I think it's time we ended things.* Here, due to the lack of deeper understanding of what PMDD really is, your lay brain makes a decision based on your limited perception and needs, that you need to end the relationship.

Context 2

Now let's run the same experiment in a different context. Imagine yourself to be a person with a hobby of reading a lot of Science books, especially those connected to the mind. You like to learn new things. Now, you get into the same situation and the same circumstances as mentioned earlier. But, here when your girlfriend tells you about her condition after the first month of outburst in your relationship, you get really curious. Previously due to your curiosity you have already learnt about the basic biology behind PMS and how it affects the female psychology. So, hearing about PMDD, you don't only feel responsible but also very intrigued to learn about it. You feel the urge to know how can such a cheerful person turn into a completely different human being, almost like a beast?

And as you start reading, everything begins to make sense. The first thing you realize is that, PMDD is nothing like the common PMS that almost every girl faces. It is the extremely violent form of PMS, which is very rare. Due to the intense hormonal storms of PMDD inside a woman's head, her cognitive reality changes drastically during the pre-menstrual days. You start to realize what your woman has to go through every month, due to her condition. A man can never even imagine in his wildest dreams how such storm feels like.

PMDD leads to the worst of hormonal mood swings. Every month during these days she turns into a completely different person filled with hopelessness and gloom. As soon as the tides change she comes back to her real cheerful self. It is this condition which makes her say the things she would never say in a lucid mental state.

You begin to understand that for most women with PMS the hormonal changes are manageable, and they are able to somehow keep their agitation to themselves. But for your special lady, the story is different and quite unmanageable. Most weeks of the month she is brainy, creative, enthusiastic, cheerful and optimistic, but a mere shift in the hormonal flood on certain days makes her absolutely hopeless about the future, about herself, about your relationship and basically about everything that she can think of. On those days her inner instability forces her to hate herself as well as get irritated at every single action you take. And the most fascinating thing about her mental state during that time is that the hopelessness caused by hormonal imbalances feels so damn real to her that she literally perceives it as the everlasting reality of her life. The utter hormonal turbulence completely transforms her cognitive reality from a cheerful one to a gloomy one. It constructs an altered state of consciousness, in which she becomes a different personality filled with nothing but hatred and rage.

She becomes absolutely blind to all the cheerful moments of her life. And she gets so restless that she explodes with insulting words towards you. Over time, you learn that the best to do in this situation is to do nothing and just be there with her. And every time she gets cranky, you simply learn to remain patient and unaffected by her words. Every month, once the hormonal storm wears off, she comes back to her original sunny state. In time, you grow more attachment for her, and she even becomes fonder of you, because you are always there for her, even when she is mad as hell. Thus, you don't ever feel to leave her, rather together you stay forever and beyond.

Analysis

In both contexts, you had two available options to choose from – *leave* or *stay*. Yet, you made totally opposite decisions in exactly the same circumstances. The only thing that was different is your understanding and experience.

In the first context, you were a layman with a general view of the world. You perceived everything in a generalized manner, with no further need of your own to explore and have deeper understanding of a phenomenon. Hence, when it came to decide whether to leave or to stay in the relationship, your brain made the decision based on your generalized understanding of everything, and you willfully preferred to leave your woman in the pursuit of a better prize, with less crankiness.

In the second context, you had a better grip over natural phenomena. Moreover, you had the curiosity to understand things, in a better way than the general public. And your understanding allowed you to try seeing the world from your woman's perspective. And the more you tried, the better you became at being next to her when she needed you the most and hated you at the same time. You got experienced at it.

Naturally, the thought of leaving her, never occurred in your mind. And even if it did, you brushed off as the spur of the moment. Because your PFC already had sufficient data on the situation to analyze and come up with a positive outcome. Your experience here served as the very foundation, of your willful support towards your woman.

Conclusion

Thus, we can conclude that simply the mental element of personal experience is capable of altering a person's willful decision. Hence, it is not about whether you have free will, rather it is about whether you have enough experience to make the best possible willful decision in the current moment of life.

Natural Sciences are all about fascinating causality. But it is not *merely* causality, it is much more than that. Causality is the amazing foundation, on which all phenomena of life and everything else take shape. And the purpose of natural sciences is to understand the mechanisms underneath causality with intricate detail.

We the scientists take pleasure in attempting to understand things that have baffled humanity since its birth. And the freedom of will in our neurobiology enables us to keep looking, never ever to stop. The continuity of our cellular processes allows our mind to evolve and be better with each passing moment.

We are a fantastic species run by biology. Biology is run by intricate cellular mechanisms. Cellular mechanisms are run by Nature. Thus, the more we attempt to understand Nature, the more we get closer to our existential properties. Nature holds the record of our past. And it also holds the key to our future.

* * *

Bibliography

- Abel T, Lattal KM. Molecular mechanisms of memory acquisition, consolidation and retrieval. Curr Opin Neurobiol 2001; 11: 180-7.
- Abram Hoffer and Humphrey Osmond, The Hallucinogens (New York: Academic Press, 1967).
- Andresen, Jensine, and Robert Forman, eds. Cognitive Models and Spiritual Maps. Bowling Green, Ohio: Imprint Academic, 2000.
- Ashbrook, James, and Carol Albright. The Humanizing Brain: Where Religion and Neuroscience Meet. Cleveland, OH: Pilgrim Press, 1997.
- Azari, Nina, Janpeter Nickel, Gilbert Wunderlich, Michael Niedeggen, Harald Hefter, Lutz Tellmann, Hans Herzog, Petra Stoerig, Dieter Birnbacher, and Rudiger Seitz. "Neural Correlates of Religious Experience." European Journal of Neuroscience 13, no. 8 (2001)

- Baars, B. (1988), A Cognitive Theory of Consciousness (New York: Cambridge University Press).
- Benfenati F, Onofri F, Giovedi S. Protein-protein interactions and protein modules in the control of neurotransmitter releas
- Belisheva, N. K., Popov, A. N., Petukhova, N. V., Pavlova, L. P., Osipov, K. S., Tkachenko, S.E., & Baranova, T.I. (1995). Quantitative and qualitative evaluations of the effect of geomagnetic variations on the functional state of the brain. Biophysics, 40
- Beauregard, Mario, and Denyse O'Leary. The Spiritual Brain. New York: HarperCollins, 2007.
- Beauregard, Mario, and Vincent Paquette. "Neural Correlates of a Mystical Experience in Carmelite Nuns." Neuroscience Letters 405, no. 3 (2006)
- Benson, Herbert. Timeless Healing: The Power and Biology of Belief. New York: Scribner, 1996
- Bogen, J.E.(1995a), 'On the neurophysiology of consciousness: Part

I. An overview', Consciousness and Cognition, 4.

- Bogen, J.E. (1995b), 'On the neurophysiology of consciousness: Part II. Constraining the semantic problem', Consciousness and Cognition, 4.
- Bremner, J. D., R. Soufer, et al. (2001). "Gender differences in cognitive and neural correlates of remembrance of emotional words." Psychopharmacol Bull 35 (3).
- Brothers, L. (2002). The social brain: A project for integrating primate behavior and neurophysiology in a new domain. In J. T. Cacioppo et al. (Eds.), Foundations in neuroscience. Cambridge, MA: MIT Press.
- Buss, D. D. (2003). Evolutionary Psychology: The New Science of Mind, 2nd ed. New York: Allyn & Bacon. Buss, D. M. (1989). "Conflict between the sexes: Strategic interference and the evocation of anger and upset." J Pers Soc Psychol 56 (5).
- Buss, D. M. (1995). "Psychological sex differences. Origins through sexual selection." Am Psychol 50 (3).

- Buss, D. M. (2002). "Review: Human Mate Guarding." Neuro Endocrinol Lett 23 (Suppl 4).
- Buss, D. M., and D. P. Schmitt (1993). "Sexual strategies theory: An evolutionary perspective on human mating." Psychol Rev 100 (2).
- Churchland, P.S. (1986), Neurophilosophy (Cambridge, MA: The MIT Press).
- Churchland, P.S. & Ramachandran, V.S. (1993), 'Filling in: Why Dennett is wrong', in Dennett and His Critics: Demystifying Mind, ed. B. Dahlbom (Oxford: Blackwell Scientific Press).
- Churchland, P.S., Ramachandran, V.S. & Sejnowski, T.J. (1994), 'A critique of pure vision', in Large- scale Neuronal Theories of the Brain, ed. C. Koch & J.L. Davis (Cambridge, MA: The MIT Press).
- Crick, F. (1994), The Astonishing Hypothesis: The Scientific Search for the Soul (New York: Simon and Schuster). Crick, F. (1996), 'Visual perception: rivalry and consciousness', Nature, 379.
- Crick, F. & Koch, C. (1992), 'The problem of consciousness', Scientific American, 267.

- d'Aquili, Eugene. "Senses of Reality in Science and Religion." Zygon 17, no 4 (1982)
- d'Aquili, Eugene. "The Biopsychological Determinants of Religious Ritual Behavior." Zygon 10, no. 1 (1975)
- d'Aquili, Eugene. "The Myth-Ritual Complex: A Biogenetic Structural Analysis." Zygon 18, no. 3 (1983)
- d'Aquili, Eugene, and Andrew Newberg. The Mystical Mind: Probing the Biology of Religious Experience. Minneapolis: Fortress Press, 1999.
- Dahlstrom A, Fuxe K (1964). "Evidence for the existence of monoamine-containing neurons in the central nervous system". Acta Physiologica Scandinavica 62.
- Daly DD. 1958. Ictal affect. Am J Psychiatry.
- Darwin, Charles. "On the origin of species by means of natural selection" (original edition, 1859).
- Darwin, Charles. "The Descent of Man" (original edition, 1871).
- Dawkins, Richard. "The God Delusion", Bantam Press, 2006

- Dennett, Daniel. Breaking the Spell: Religion as a Natural Phenomenon. New York: Penguin, 2007.
- Dennett, D.C. (1991), Consciousness Explained (Boston, MA: Little, Brown and Co.).
- Dewhurst, Kenneth, and A. W. Beard. "Sudden Religious Conversions in Temporal Lobe Epilepsy." British Journal of Psychiatry 117 (1970)
- Dixson, B. J., & Brooks, R. C. (2013) "The role of facial hair in women's perceptions of men's attractiveness, health, masculinity and parenting abilities". Evolution and Human Behavior 34, 236-41
- Dewhurst K, Beard AW. Sudden religious conversions in temporal lobe epilepsy. 1970 Epilepsy Behav 2003
- Devinsky O, Lai G. Spirituality and religion in epilepsy. Epilepsy Behav 2008.
- Devinsky, O., Morrell, MJ, Vogt, BA. (1995) 'Contribution of anterior cingulate cortex to behavior', Brain, 118.
- Downing PE, Jiang Y, Shuman M, Kanwisher N. 2001. A cortical area selective for visual processing of the human body. Science.

- Esquirol, Étienne 1845. Mental maladies; a treatise on insanity (original French edition 1838).
- Ellis, E., Ames, M.A., 1987. Neurohormonal functioning and sexual orientation: a theory of homosexuality–heterosexuality. Psychol. Bull. 101.
- Falk, D. et al. Early hominid brain evolution: a new look at old endocasts. Journal of Human Evolution 38, (2000)
- Farah, M.J. (1989), 'The neural basis of mental imagery', Trends in Neurosciences, 10.
- Finlay BL, Darlington RB (1995) Linked regularities in the development and evolution of mammalian brains. Science 268.
- Freud, S. (1920) "Beyond the Pleasure Principle"
- Frith, C.D. & Dolan, R.J. (1997), 'Abnormal beliefs: Delusions and memory', Paper presented at the May, 1997, Harvard Conference on Memory and Belief.
- Gay, Volney, ed. Neuroscience and Religion. Plymouth, UK: Lexington Books, 2009.
- Gazzaniga, M. S. (1985). The social brain. New York: Basic Books.

- Gazzaniga, M.S. (1993), 'Brain mechanisms and conscious experience', Ciba Foundation Symposium, 174.
- Geschwind N. "Behavioural changes in temporal lobe epilepsy". Psychol Med. 1979.
- Gilbert SL, Dobyns WB, Lahn BT (2005) Genetic links between brain development and brain evolution. Nat Rev Genet 6.
- Gloor, P. (1992), 'Amygdala and temporal lobe epilepsy', in The Amygdala: Neurobiological Aspects of Emotion, Memory and Mental Dysfunction, ed J.P. Aggleton (New York: Wiley-Liss).
- Greenspan, S. I. and S. G. Shanker (2004). The first idea: How symbols, language, and intelligence evolved from our early primate ancestors to modern humans. Cambridge, MA: Da Capo Press.
- Grady, D. (1993), 'The vision thing: Mainly in the brain', Discover, June.
- Grüsser OJ, Landis T. 1991. The splitting of "I" and "me": heautoscopy and related phenomena. In: Visual agnosias and other disturbances of visual

perception and cognition. Amsterdam: MacMillan.

- Hall, Daniel, Keith Meador, and Harold Koenig. "Measuring Religiousness in Health Research: Review and Critique." Journal of Religion and Health 47, no. 2 (2008)
- Harris, Sam, Jonas Kaplan, Ashley Curiel, Susan Bookheimer, Marco Iacoboni, and Mark Cohen. "The Neural Correlates of Religious and Nonreligious Belief." PLoS One 4, no. 10 (October 1, 2009)
- Halgren, E. (1992), 'Emotional neurophysiology of the amygdala within the context of human cognition', in The Amygdala: Neurobiological Aspects of Emotion, Memory and Mental Dysfunction, ed J.P. Aggleton (New York: Wiley-Liss).
- Halligan PW, Fink GR, Marshal JC, Vallar G. 2003. Spatial cognition: evidence from visual neglect. Trends Cogn Sci.
- Handbook of Emotions, Edited by Michael Lewis, Jeannette M. Haviland-Jones, and Lisa Feldman Barrett, The Guilford Press; 3rd edition (2010).

- Horgan, J. (1994), 'Can science explain consciousness?', Scientific American, 271.
- Holloway RL, Broadfield DC, Yuan MS (2004) The human fossil record, vol 3, Brain endocasts: the paleoneurological evidence. Wiley, New York
- Holloway RL (1996) Evolution of the human brain. In: Lock A, Peters CR (eds) Handbook of human symbolic evolution. Oxford University Press, Oxford
- Hebb, Donald, O. The Organization of Behavior: A Neuropsychological Theory, 1949
- Hilgard E. R., D. G Marquis. Hilgard and Marquis' Conditioning and learning.. New York, Appleton-Century-Crofts, 1961.
- Humphrey, N. (1993), A History of the Mind (London: Vintage).
- Kandel, E. R. In Search of Memory: The Emergence of a New Science of Mind, W. W. Norton & Company (2007).
- Kandel E. R. Schwartz JH, Jessel TM. Principles of neural sciences. New York; McGraw Hill, 2000.
- Kanizsa, G. (1979), Organization In Vision (New York: Praeger).

- Kinsbourne, M. (1995), 'The intralaminar thalamic nucleii', Consciousness and Cognition, 4.
- Kjaer, Troels, Camilla Bertelsen, Paola Piccini, David Brooks, Jorgen Alving, and Hans Lou. "Increased Dopamine Tone during Meditation- Induced Change of Consciousness." Cognitive Brain Research 13, no. 2 (April 2002)
- Kölmel HW. 1985. Complex visual hallucinations in the hemianopic field. J Neurol Neurosurg Psychiatry.
- Koenig, Harold. "Research on Religion, Spirituality, and Mental Health: A Review." Canadian Journal of Psychiatry 54, no. 5 (May 2009)
- Koenig, Harold, ed. Handbook of Religion and Mental Health. San Diego, CA: Academic Press, 1998
- Kuypers HGJM (1958) Corticobulbar connections to the pons and lower brainstem in man. Brain 81
- Lauglin, Charles, John McManus, and Eugene d'Aquili. Brain, Symbol, and Experience. 2nd ed. New York: Columbia University Press, 1992
- Lakoff, G. and M. Johnson (1999). Philosophy in the flesh. Basic Books: New York.

- LeDoux, J. E. (1996). The emotional brain. New York: Simon & Schuster.
- LeDoux, J.E. (1992), 'Emotion and the amygdala', in The Amygdala: Neurobiological Aspects of Emo- tion, Memory and Mental Dysfunction, ed J.P. Aggleton (New York: Wiley-Liss).
- MacKay, D.M. (1969), Information, Mechanism and Meaning (Cambridge, MA: The MIT Press).
- Matynia A, Kushner SA, Silva AJ. Genetic approaches to molecular and cellular cognition: a focus on LTP and learning and memory. Annu Rev Genet 2002; 36.
- Metzinger T. 2003. Being no one. Cambridge (MA): MIT Press.
- Miller S, Mayford M. Cellular and molecular mechanisms of memory: the LTP connection. Curr Opin Genet Dev 1999, 9.
- Milner, A.D. & Goodale, M.A. (1995), The Visual Brain In Action (Oxford: Oxford University Press).
- Nagel, T. (1974), 'What is it like to be a bat?', Philosophical Review, 83.
- Naskar, A. Homo: A Brief History of Consciousness, 2015.

- Naskar, A. Autobiography of God: Biopsy of A Cognitive Reality, 2016.
- Naskar, A. Biopsy of Religions: Neuroanalysis towards Universal Tolerance, 2016.
- Newberg, Andrew, and Jeremy Iversen. "The Neural Basis of the Complex Mental Task of Meditation: Neurotransmitter and Neurochemical Considerations." Medical Hypotheses 61, no. 2 (2003).
- Newberg, Andrew, and Mark Waldman. How God Changes Your Brain, New York: Ballantine, 2010.
- Newberg, Andrew. "How God Changes Your Brain: An Introduction to Jewish Neurotheology", CCAR Journal: The Reform Jewish Quarterly, Winter 2016.
- Newberg, Andrew, and Stephanie Newberg. "A Neuropsychological Perspective on Spiritual Development." In Handbook of Spiritual Development in Childhood and Adolescence, edited by Eugene Roehlkepartain, Pamela King, Linda Wagener, and Peter Benson. London: Sage Publications, Inc., 2005
- Newberg, Andrew. "The Neurotheology Link An Intersection Between Spirituality and Health", Alternative

and Complimentary Therapies, Vol 21 No 1, February 2015.

- Newberg, Andrew, Nancy Wintering, Dharma Khalsa, Hannah Roggenkamp, and Mark Waldman. "Meditation Effects on Cognitive Function and Cerebral Blood Flow in Subjects with Memory Loss: A Preliminary Study." Journal of Alzheimer's Disease 20, no. 2 (2010)
- Nash, M. (1995), 'Glimpses of the mind', Time.
- Papez, J. W. (1937). A proposed mechanism of emotion. Archives of Neurology and Psychiatry, 79.
- Pare, D., & Smith, Y. (1998). Intrinsic circuitry of the amygdaloid complex: Common principles of organization in rats and cats. Trends in Neurosciences, 21.
- Penfield, W.P. & Perot, P. (1963), 'The brain's record of auditory and visual experience: a final summary and discussion', Brain, 86.
- Penrose, R. (1994), Shadows of the Mind (Oxford: Oxford University Press).
- Penrose, R. (1989), The Emperor's New Mind: Concerning Computers, Minds

and The Laws of Physics (Oxford: Oxford University Press).

- P. S. de Laplace. Essai Philosophique sur les Probabilites [1814], in Academy des Sciences, Oeuvres Complotes de Laplace, Vol. 7, Gauthier-Villars, Paris (1886).
- Posner, M.I. & Raichle, M.E. (1994), Frames of Mind (New York: Scientific American Library).
- Ramachandran, V.S. (1993), 'Filling in gaps in logic: Some comments on Dennett', Consciousness and Cognition, 2.
- Ramachandran,V.S. (1995a),'Filling in gaps in logic: Repy to Durginetal.',Perception, 24.
- Ramachandran, V.S. (1995b), 'Perceptual correlates of neural plasticity', in Early Vision and Beyond, ed. T.V. Papathomas, C. Chubb, A. Gorea and E. Kowler (Cambridge, MA: The MIT Press).
- Ramachandran, V.S. and Blakeslee, S. (1999), Phantoms in the Brain: Probing the Mysteries of the Human Mind (William Morrow Paperbacks).
- Ramachandran, V.S. A Brief Tour of Human Consciousness: From Impostor

Poodles to Purple Numbers, Pi Press 2005.

- Ramón y Cajal, Santiago (1999) [1897]. Advice for a Young Investigator. Translated by Neely Swanson and Larry W. Swanson. Cambridge: MIT Press. ISBN 0-262-68150-1.
- Scoville, W. B., & Milner, B. (1957). "Loss of recent memory after bilateral hippocampal lesions". Journal of Neurology, Neurosurgery, and Psychiatry, 20.
- Schultes, Richard, Albert, Hofmann, and Christian Rätsch. Plants of the Gods: Their Sacred, Healing, and Hallucinogenic Powers. Rochester, VT: Healing Arts Press, 2001
- Searle, John R. (1980), 'Minds, brains, and programs', Behavioral and Brain Sciences, 3.
- Searle, John R. (1992), The Rediscovery of the Mind (Cambridge, MA: The MIT Press).
- Shermer, Michael. The Believing Brain: From Ghosts and Gods to Politics and Conspiracies—How We Construct Beliefs and Reinforce Them as Truths. New York: Times Books, 2011

- Silva AJ, Josselyn SA. The molecules of forgetfulness. Nature 2002; 418: 929-930.
- Steven A. Barker, John A. Monti, and Samuel T. Christian, "N,N-Dimethyltryptamine: An Endogenous Hallucinogen," International Review of Neurobiology 22 (1981)
- Streeter, Chris, J. Eric Jensen, Ruth Perlmutter, Howard Cabral, Hua Tian, Devin Terhune, Domenic Ciraulo, and Perry Renshaw. "Yoga Asana Sessions Increase Brain GABA Levels: A Pilot Study." Journal of Alternative and Complementary Medicine 13, no. 4 (May 2007).
- Strassman, R. "DMT: The Spirit Molecule" 2001.
- Tononi, G., & Edelman, G. E. (1998). Consciousness and complexity. Science, 282.
- Tulving, E. (1983), Elements of Episodic Memory (Oxford: Clarendon Press).
- Turner, J. H. (2000b). On the origins of human emotions: A sociological inquiry into the evolution of human affect. Stanford, California: Stanford University Press.

CPSIA information can be obtained at www.ICGtesting.com
Printed in the USA
LVOW11s2337140716

496331LV00047B/9/P

9 781535 080552